# PHILOSOPY OF CONSCIOUSNESS

## A Draft

Pavel Ivanov

Trafford Publishing
www.trafford.com

Ivanov, Pavel B.
Philosophy of Consciousness. A Draft

Order this book online at www.trafford.com
or email orders@trafford.com

Most Trafford titles are also available at major online book retailers.

© Copyright 2009 Pavel B. Ivanov.
Email: unism@ya.ru

All rights reserved. No part of this publication may be reproduced, stored in a retrieval system, or transmitted, in any form or by any means, electronic, mechanical, photocopying, recording, or otherwise, without the written prior permission of the author.

Note for Librarians:
A cataloguing record for this book is available from Library and Archives Canada at www.collectionscanada.ca/amicus/index-e.html

Printed in Victoria, BC, Canada.

ISBN: 978-1-4269-0245-1 (sc)

ISBN: 978-1-4269-0790-6 (dj)

ISBN: 978-1-4269-1098-2 (e-book)

We at Trafford believe that it is the responsibility of us all, as both individuals and corporations, to make choices that are environmentally and socially sound. You, in turn, are supporting this responsible conduct each time you purchase a Trafford book, or make use of our publishing services. To find out how you are helping, please visit www.trafford.com/responsiblepublishing.html

Our mission is to efficiently provide the world's finest, most comprehensive book publishing service, enabling every author to experience success. To find out how to publish your book, your way, and have it available worldwide, visit us online at www.trafford.com

Trafford rev. 6/26/2009

 www.trafford.com

North America & international
toll-free: 1 888 232 4444 (USA & Canada)
phone: 250 383 6864 • fax: 250 383 6804 • email: info@trafford.com

The United Kingdom & Europe
phone: +44 (0)1865 487 395 • local rate: 0845 230 9601
facsimile: +44 (0)1865 481 507 • email: info.uk@trafford.com

# Contents

PREFACE .................................................................................. 1

HIERARCHICAL APPROACH ................................................. 5
    Hierarchical Integrity ........................................................... 8
        Structures ..................................................................... 11
        Systems ........................................................................ 14
        Hierarchies .................................................................. 19
    Hierarchical Conversion ................................................... 24
    Fundamental Principles .................................................... 29

LOGIC ..................................................................................... 33
    Levels of Logic ................................................................. 35
        Syncretic, analytical and synthetic logic ..................... 35
        Metaphysics, dialectics and diathetics ........................ 36
        Levels of intelligence .................................................. 38
        Regularity and truth .................................................... 39
        Discreteness and continuity ........................................ 40
        Organization in logic ................................................... 41
    Classical Logic ................................................................. 42
        What is classical? ........................................................ 42
        Branches of classical logic .......................................... 43
        Logical forms .............................................................. 44
        Adequacy, truth, correctness ....................................... 45
        Fundamental principles ............................................... 46
        Fallacies ...................................................................... 48
    Dialectical Logic .............................................................. 49
        What is dialectics? ...................................................... 50
        The origin of dialectics ............................................... 51
        Logical forms .............................................................. 52
        Fundamental principles ............................................... 53

Diathetical Logic ........................................................................ 56
    Why diathetics? ................................................................... 57
    Logical forms ...................................................................... 57
    Fundamental principles ...................................................... 60

Formal Logic ............................................................................. 63
    Classical logic .................................................................... 64
    Logic and mathematics ...................................................... 67
    The basics of diathetical logic ........................................... 71

## ONTOLOGY OF CONSCIOUSNESS ................................. 85

Ontological Roots of Subjectivity ........................................... 86
    The central problem of philosophy ................................... 88
    Levels of reflection ............................................................ 92
    Subjectivity as universal mediation ................................... 94
    The subject and activity ................................................... 101

Inside the Subject .................................................................. 108
    Unfolding mediation ........................................................ 108
    Mental processes .............................................................. 116
    From awareness to consciousness ................................... 125
    Ideation ............................................................................ 127

Hierarchy of Consciousness .................................................. 134
    Philosophical categories .................................................. 135
    Reflection schemes .......................................................... 137
    Levels of subjectivity ...................................................... 138
    Levels of culture .............................................................. 139
    Levels of spirituality ........................................................ 140

## EPISTEMOLOGY OF CONSCIOUSNESS ...................... 147

Hierarchical Methodology ..................................................... 152
    Between the object and the product ................................ 153
    Observing the subject ...................................................... 159
    Extrasensory reflection? .................................................. 163
    Theories of consciousness ............................................... 164

Cultural dependence of science ............................................. 168

Scheme Transfer: Physical Psychology ................................ 173
    Observer in physics ......................................................... 175

Physical methods in psychology ............................................. 177
The foundations of physical psychology .............................. 181
The scheme of Newtonian mechanics .................................. 183
Motivation and temperament ................................................ 186

# CONSCIOUSNESS IN PSYCHOLOGY ................................. 191

General Psychology ..................................................................... 196
    Extent hierarchy ................................................................. 197
    Meaning and sense ............................................................ 200
    Psychological sets ............................................................. 202
    Psychological dimensions of activity ............................ 203
    Schemes of activity ........................................................... 207

Differential psychology ............................................................. 211
    Hierarchy of personality .................................................. 214

Social Psychology ....................................................................... 215
    Transactional analysis ...................................................... 217
    Collective behavior ........................................................... 221

Animal psychology ..................................................................... 222

# ASPECTS OF CONSCIOUSNESS ............................................ 225

Consciousness and Physiology ................................................. 226
    Mind and body ................................................................... 226
    Consciousness and the brain ........................................... 231
    Localization of mental processes ................................... 233

Space and Time .......................................................................... 236

Development of Consciousness ............................................... 239

Education .................................................................................... 243

Consciousness and Communication ....................................... 246

Altered States of Consciousness .............................................. 253

Psychotherapy ............................................................................ 255

# ETHICS OF CONSCIOUSNESS ............................................... 265

Syncretic Ethics .......................................................................... 267
    Mediation ............................................................................ 267

| | |
|---|---|
| Universality | 268 |
| Analytical Ethics | 271 |
|     Awareness | 272 |
|     Intentionality | 274 |
|     Responsibility | 275 |
| Synthetic Ethics | 278 |
|     Conscience and esteem | 278 |
|     Immortality | 280 |

# PREFACE

Many humans have a persistent habit of thinking about themselves as of a special kind of living creatures, distinguished from all the other beings by possessing something they call *consciousness*. There are different views on the adequacy of this idea, from complete denial of its validity to considering consciousness as prior to any existence at all. While the both extremes lead to the impossibility of scientific study, the major part of researchers prefer to be pragmatic and adopt "technical" approaches of various kinds, dealing with special models of consciousness that incorporate some of its aspects related to the particular science. Obviously this can only be a temporary solution, and the fundamental questions concerning the nature of consciousness arise with more vigor whenever the development of a research technique visibly approaches the limits of its applicability. One needs a unified view of consciousness, providing a common frame for all the special sciences that investigate particular manifestations of consciousness, and thus avoiding misunderstanding due to terminological confusion and methodological incompatibility. This is what philosophy of consciousness is expected to provide.

There are many ways of unfolding philosophy of consciousness: some of them focus on the methodology of science, some others consider it from an aesthetic or ethic angle. A unified view will include all these aspects as the possible presentations of the same.

Scientific study of consciousness is essentially analytical, since it separates "consciousness" from "not consciousness," as opposing concepts. Such division can only be established in a limited domain, and there are questions that cannot be answered by science at all. For instance, no science can tell how it feels to be conscious, or how a conscious being ought to behave. These are not scientific problems, and their consideration in science would lead to their abstract re-formulations answering to questions quite different from original and valid only within the limits of applicability of the special science. It is philosophy of consciousness that is to synthesize the partial models of special sciences, as well as non-scientific views, into a general picture, demonstrating how all the possible descriptions refer to different aspects of the same phenomenon, and thus indicating what is relevant in a particular approach, or what lies beyond its scope.

Philosophy is not science. It does not "study" anything; it only discovers the possible directions of development for special and general sciences. Philosophers must consider scientific results—but science will never remain their only source of ideas. Every area of human culture can influence the development of philosophy, and receive creative impulses in return.

Any philosophy develops its own categories, explicitly or implicitly connecting them to each other in categorial schemes. While philosophical categories may sometimes be denoted with the common words, these words acquire specific connotations in the context of a particular philosophy and their meaning is to be derived from their place in the whole, rather than from the previous experience of the reader, and a correct definition of the fundamental terms (*e.g.* "life", "activity", or "consciousness") only follows from their usage in a variety of contexts. The language of this book is different from that of any other philosophical or scientific work, albeit resembling them in certain respects. The abstract schemes of my philosophy have nothing to do with mathematics, though one may be tempted to identify some of them with the common mathematical constructs. Even the English language cannot be of much help, since my word usage may be far from commonly accepted, and the ways of expression may even deviate from the grammatical norm.

If so, who could be interested in a book like that? It is certainly not for those who are happy within their profession and do not need any thoughts that cannot be converted in hard currency. However, there are many people who do not much rely on efficient professionalism and seek for an integrative view equally applicable to any special area, as well as outside any specialty at all. For those my philosophy of consciousness may happen to be of some use—at least as exercise in an uncommon logic and a vaccine against mental stagnation.

To comprehend what consciousness is, it is necessary to determine the place of consciousness and subjectivity in the hierarchy of the world (ontology), explain how conscious beings can comprehend themselves (epistemology), and indicate how their being conscious influences their behavior and their internal organization (ethics). In Hegelian terms, we are first interested in what consciousness is "in itself," then we consider how it looks "for itself," and we have to complete our inquiry investigating how consciousness "for itself" could arise from what it is "in itself," and conversely, how consciousness grows in the process of self-comprehension and self-determination. Omission of any of these elements would make consideration essentially incomplete and hence unsatisfactory, provoking people to seek for other possible explanations.

However, considering the three aspects of consciousness in philosophy is different from the analytical approach of science, which tries to separate the different aspects of the same thing, to independently treat them in different sciences. Each special science is relatively closed; its basic postulates are thought to be of the same kind as the results derived, without formal intersection with any other theories. Attempts to reproduce this artificial isolation in philosophy have adversely influenced philosophical literature, making it too dependent on special sciences instead of guiding science suggesting it fundamental methodological principles. Without a sound philosophy, science is bound to produce a lot of confusion, which is re-inherited by scientifically-minded philosophy, thus entirely obscuring things.

However, the idea of universal inter-connection has never died, and many scientists think about different sciences as complementary descriptions of the same. Holistic ideas are becoming ubiquitous. Scientists invent ever new "boundary" disciplines, "meta-sciences", or universal paradigms. Still, the level of integrity that can be achieved within science does not match the integrity of the Universe as revealed in our practical life. This is especially so for comprehension of consciousness, and many difficulties encountered by earlier researchers were due to their original analytical attitude as imposed by the dominant cultural standards.

My philosophy of consciousness is based on a general hierarchical approach, including a formal derivation scheme that I call diathetical logic. One can hardly describe hierarchies without a special logical apparatus, since, as it is well known, classical logic encounters serious difficulties treating motion and development, and numerous alternative logics appear as an attempt to cope with inherent contradictions in modern mathematics. A brief summary of hierarchical approach is given in the first chapter; the next chapter outlines its application to logic. Those who are not much inclined to abstractions can skip then and proceed directly to ontology of consciousness, considering any formal schemes as mere graphical expression for the non-technical explanations accompanying them.

My approach continues the materialist tradition in consciousness studies going back to the early works of Karl Marx and applied to numerous theoretical and practical problems by many researchers in the former USSR (E. V. Ilyenkov, L. S. Vygotsky, A. N. Leontiev, A. I. Mescheryakov, S. L. Rubinstein, B. M. Teplov and others). Such Russian scientists as Ivan Sechenov, Vladimir Bekhterev, Ivan Pavlov and Pyotr Anokhin contributed to the materialist understanding of consciousness, along the natural-scientific lines. However, modern materialism is far from considering consciousness as mere mechanical product; the differences between the main philosophical schools are to be discussed in brief. However, analyzing the position of any particular writer, or tracing parallels between my views and the ideas of the others, is outside the scope of this book; I am sure that the reader will easily find more details in the available literature.

Since I consider consciousness on the basis of a definite philosophical position, I will have to indicate, which directions of consciousness research are compatible with that stand, and what contradicts it. It is only in this sense that I say how consciousness must be studied, and how it cannot be studied. In this draft of philosophy, I do not give final answers—I only suggest topics for discussion and present an ideological stand, which is far from being commonly accepted; I must apologize if somebody's feelings are hurt.

Why a draft? Well, there are a few solid reasons. Serious changes are underway in the world, and I find that it is high time for my philosophy of consciousness to contribute to that development; that is why I cannot wait until the problems discussed in this book receive a more profound and comprehensive consideration. Probably, such a solid treatise would be much lengthier and less friendly to the reader; the present volume seems to be enough for a general introduction. Finally, any individual effort can only outline the approach and prompt the readers to proceed in the indicated direction on their own. No philosopher can cover the whole range of problems related to reason and subjectivity, and hence any philosophy of consciousness is bound to always remain a draft.

# HIERARCHICAL APPROACH

Ideas do not significantly depend on the language. One can express them differently, using alternative formulations, inventing peculiar terms or employing cumbrous phrases and formulas, talking science-like or common-sense... Sometimes ideas can be expressed without any words at all, by means of practical example and human sympathy. Still, in many cases we have to put our thoughts in words and hence depend on their capability to convey ideas.

Making science, we do not much care for terminology, and scientists often borrow their terms from everyday language, filling them with a content that has nothing in common with the original meaning. For instance, physicists speak of quark flavors—while everybody knows that quarks do not smell, and the term should not be understood literally. Sometimes, especially in social sciences, scientific terms sound too like the ordinary language, and a mental effort is needed to abstract from the common usage and come to the specific connotation assumed by the particular science. Thus, the terms "space" and "time" in a physical theory should not be confused with the real space and time; physicists deal with a single aspect of spatial and temporal relations, providing physical models that do not describe their prototype in general. Similarly, "value" in economy is not the same as social or personal value, and a logical "truth" is not necessarily true in real life.

In philosophy, things get even worse. Since philosophers discuss universal categories rather than specific regularities, it is almost impossible to find a single word to denote a philosophical category; the same category may appear in different contexts under different names, and a slight change in wording may mean a drastically new viewpoint. One has to derive the meaning of a philosophical text from its whole, or even from a selection of texts by the same author, or a group of authors.

In this situation, one could only dream of a standard instrument for developing a philosophy, and a uniform language for its presentation. Many people tried to invent such a common platform for philosophy. However, all such attempts suffered from the same reductionism disease: they took an artistic method, or scientific paradigm, and pretended to discover a universal core of philosophy. Though many of these metaphors made valuable contributions to philosophical thought, they failed to fulfill their declared task, since the correct direction is exactly the opposite, and it is philosophy's task, to provide conceptual frameworks to science and art, rather than the reverse.

I suggest what I call a "hierarchical approach" as yet another effort on the way to a unified philosophy, and I admit that better doctrines can be built in the future. Still, this particular mental instrument has been helpful to unfold my philosophy of consciousness, and conveniently organize this book. Probably, it could also come handy for something else.

Why "hierarchy"?

The word is of Greek origin, and it can be approximately translated as "sacred subordination". The term has been primarily introduced by Christian scholars[1] to describe the way God has arranged all the entities. The centralized organization of the church was said to mimic the organization of celestial beings (archangels, angels *etc*); its secular analog was found in absolute monarchy. Medieval theology was essentially static, it did not need the idea of change and development, since the order admittedly brought to things by God was thought to be already perfect. The word "hierarchy" has retained a strong structural connotation in the modern language as well, denoting mainly tiered structures, rigid sets of pre-defined levels, with fixed relations between them. This does not satisfy a thoughtful philosopher, since the levels of such hierarchy are separated from each other, with no "vertical" dynamics, and the source of order remains a mystery.

Nevertheless, the idea of universal natural order is very attractive. We are usually aware of some "subordination" of things in nature and human culture; there are processes of very different scales; growth and evolution are all around us—but we also see that changing things still preserve their

---

[1] The idea of hierarchy is much older, with its origin in the mythological cosmology of the first primitive societies. Since the relations between the levels of thus pictured cosmos were unknown, they seemed to be imposed by some supreme force, deity, and such an order was correctly called "sacred"—hierarchy.

integrity, and all the levels belong to the same something. In 1982, I have suggested the word "hierarchy" to denote this hard-to-grasp natural law, which is much more flexible than the old "sacred order" as imposed by heaven once and for ever.

Probably, one should rather invent a special word for the kind of order that can be reversed at any moment and exists only in a relative way, in a specific situation. Examples of such linguistic exercises can be found in the literature (*e.g.* "heterarchies" of E. Eliseyev). Quite often, the specificity of the idea was attributed to some other categories (like "structure", "system", "integrity", "totality", and many other term usually borrowed from special sciences). Personally, I would prefer a neologism "idiarchy", which could be translated as "the own order of things", from Greek "*idios*" (own) and "*arhe*" (order, dominance). This term stresses both the universality of ordering, as well as its individuality. However, such artificial words do not usually help a reader to catch the author's ideas, and that is why I have decided to keep, in this book, the old name "hierarchy" (occasionally mentioning "idiarchy" as a full synonym) and thus preserve the fundamental connotation of something with multiple levels joined by relations of domination and control. Comparing the category "hierarchy" with other categories, I will try to reveal its specific content.

This linguistic trick will supposedly make hierarchies quite intuitive in hierarchical structures and systems. It also draws in the usual idea of time as irreversible one-directional order (unlike time coordinate in physics), and the popular conception of ontological development reproducing the basic features of phylogenic development. Common phrases like "to raise to a new level", applied to developing entities, carry the same hierarchical load. In my understanding of hierarchies, order still remains "sacred"—in the sense of all-penetrating universality of the principles governing the development of the world. But it should not be deified, mystified, put "beyond", or "above", real things rather than in their nature.

Of course, I do not pretend to invent everything on my own. Any portion of the hierarchical approach could be found in the literature, starting from cuneiform inscriptions of Ancient Mesopotamia up to the most recent multimedia books. It may be strange and a little embarrassing to observe how many people still do not grasp the simplicity of hierarchical ideas and invent cumbrous and clumsy conceptualizations to explain that, which obviously follows from the hierarchical approach. Everything is ready for the whole, but the minds are not yet flexible enough to put together the scattered pieces.

I consider this book as yet another intellectual exercise aimed to increasing the universality of human thought.

## Hierarchical Integrity

Hierarchical approach naturally continues the historical line of understanding complexity. The end of the XIX century put forward structuralism, and everything was said to be a structure. In the second half of the XX century, such static descriptions were already felt to be insufficient, and the notion of a system came to account for regular transformation of structures; however, systemic motion is not enough to explain development, but the first attempts to include it in consideration only added tiered architecture to structures and systems, thus producing hierarchical structures and hierarchical systems. The typical problem with this approach was that nobody could say where the multiple levels came from, and therefore hierarchies had to be postulated, thus becoming rigid abstractions, rather than developing entities. Things become much more logical if one suggests that hierarchy is something different from systemic organization, or structure, and that the levels of a hierarchy represent the history of its development. In this sense, one can speak of hierarchical structures as imprints of the object's development on its internal organization, while hierarchical systems reflect the dependence of the object's functionality on its natural history. Hierarchical approach thus becomes a synthesis of structuralism and systems theory.

One could formally derive hierarchy (or, rather, idiarchy) from the idea of *integrity*.

Speaking of something, we first find it as distinguished from the rest of the world, as something uniquely individual; on this level, the thing's integrity means just being itself, isolated from other things. Since no internal organization or external relations are considered, this syncretic type of integrity could be called *simplicity*. Not much can be inferred from such a primitive integrity—still, this is the necessary first stage of any study, the recognition of the problem.

On the next level, simplicity gives way to observation of external dependencies and internal non-homogeneity, and we are interested in what makes the thing what it is. Here, we come to considering *complexity*, arising from both the thing's interaction with its environment producing specific structures, processes, or kinds of development, as inherent to the thing, which is no longer unique and simple and the peculiarity of its different aspects feeds many special sciences. In complex things, integrity may seem violated, being potential rather than actual, and a "metascientific" approach is required to provide a unified view.

The level of *unity* restores simplicity retaining complexity. The thing becomes entirely reflected in its environment, while this environment is completely represented "inside" the thing. Any feature of the thing corresponds to a line in its history; any behavior is non-local, being controlled by some higher-level development. Science can never deal with unity, it is only in practical activity that we comprehend it.

Now, let us look closer at complexity. On its lowest level, one founds many simple things instead of a single thing, and the impression of complexity comes from the immense number of the instances of simplicity, rather than from any distinctions, or intricate motion. Since each member of this collection is simple, they are all unique and different from each other; on the other hand, they are unrelated and hence indistinguishable. This kind of complexity can be named *multiplicity*.

When different things are somehow related to each other, we come to a higher level of complexity, *organization*. The interrelated individual things are no longer simple; however, related to the whole, they retain some singularity becoming its *elements*. On this level, elements are always considered as opposed to organization.

Of course, there can be partially organized multiplicity, or multiplied organization. For instance, one can consider classes of entities, and multitude of classes instead of a cloud of individual entities. The members of the same class are still undistinguishable, and any of them can represent the whole class, and conversely, a class can be derived from its arbitrary member. Therefore, the complexity of such classes is related to the number of its elements, and the classes can be ordered by their cardinal numbers. Obviously, this two-level picture can be extended to any number of levels. Similarly, one can consider a non-structured multiplicity of highly organized entities, and a cardinal hierarchy induced by their inner organization. In this way, multiplicity and organization penetrate each other, producing complexity of the next level, *order*.

Here, like in the common language, the word "order" has the connotations of both "well arranged" (strongly coupled) and "properly made" (corresponding to its place in the world). In the historical perspective, the difference between multiplicity and order reproduces the ancient opposition of Chaos and Cosmos—the opposition that was thought to give birth to all the earthly things. The earthly way from Chaos (multiplicity) to Cosmos (order) is readily associated with the intermediate level of complexity that I call organization, which introduces some congruence into chaotic multiplicity, while leaving enough space for extensive and intensive development of local order. That is multiplicity means disorder, organization brings partial order, while the highest level of complexity assumes complete, universal order.

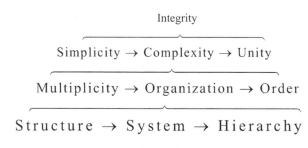

Figure 1. The hierarchy of integrity.

Using the same logic, we can distinguish three levels of organization, thus coming to structure, system and hierarchy.

The first category of this triad, *structure*, refers to internal coherence, representing the object as a collection of elements and their links. This representation is least different from multiplicity, the only new feature being the division of the multiplicity into two classes, one called "elements" and the other called "links". Being the internal characteristic of the object, structure may be thought of as the *static* aspect of the object.

The inverse of structure is *system*, the second level of coherence. It refers primarily to the external manifestations of the object, the way it "moves" in its outer space, altering its relations with the environment. Since these relations are somehow structured, system may be generally considered as the way of transforming one structure into another. So, the basic category at the systemic level is "transformation", or "transition"—and therefore system represents the object's *dynamics*.

Logically, the next level of coherence should be the synthesis of the internal description provided by structure and the external systemic treatment. It should consider the object both statically and dynamically, so that systemic transformations lead to the internal changes in the object, which nevertheless retains some of its structural features as to remain *the same* in these transformations. This is the level of *development*—and the synthesis of structural and systemic features is *hierarchy*.

Thus, complexity itself becomes complex, comprising the hierarchy of possible forms (figure 1). One level of distinction provides the triad of multiplicity, coherence and order—on another level, one might distinguish structural complexity, systemic (functional) complexity, and hierarchical (developmental) complexity. Incidentally, this sequence reflects the history of methodological thought in the XX century: the beginning of the century was marked by the structural approach, which gave way to systemic approach in the middle of the century, while the end of the XX century passed under the dominance of the idea of development, which receives its formal expression in the hierarchical approach, which gives a clear criterion for distinguishing structures, systems and hierarchies, or rather structural, systemic and hierarchical aspects in the same thing.

## *Structures*

Speaking of structure, we usually mean some internal organization of something, the arrangement of its parts. This is exactly the meaning of the Roman word *structura* (from *struere*—to build; the same root can be found in another related word: *construction*). We say that there are distinct *elements*, connected with a number of *links*, or *relations*[2]. Thus connected elements are called linked, or related to each other. The quality of elements and links depends on the nature of the thing, whose structure is considered, as well as on the level of consideration.

In general, some elements are linked in the structure, while some are not. In the limit case, when there are no links at all, we come to mere multiplicity; in the opposite limit, every element of the structure is directly linked to every other element (closely coupled structures).

However, structure, as a level in the hierarchy of integrity, is more than

---

[2] Structures are often mathematically modeled as abstract sets with a number of relations defined on them: the elements of the set represent the elements of the structure, while the links are associated with the *n*-tuples of the elements belonging to an *n*-place relation. The support set may be either discrete, or continuous, or of a higher cardinality. Accordingly, the collection of relations will vary from the finite number of element pairs to connectivities on a non-trivial manifold. Links can be either rigid, or stochastic, or any combination of the two cases. All these possibilities are in the scope of traditional mathematics, which could, in general, be called science about structures.

its elements and links, being a whole. Understanding this integrity requires considering *induced hierarchical structures*, which could be derived as follows.

In addition to *direct* links between the structure's elements, one can consider *indirect* links. For instance element $a$ is linked to element $b$ ($a \to b$), but not linked to element $c$; however, element $b$ is linked to element $c$ ($b \to c$). In the mathematical slang, such structures are called intransitive. Nevertheless, we can always consider the combination of the two direct links: $a \to b \to c$. In this scheme, element $b$ is said to *mediate* the relation between $a$ and $c$. There can be many other elements mediating this relation; also, one can construct longer chains of mediation, with two, three or more mediating elements between $a$ and $c$. Collecting all possible mediations, we obtain an indirect link $a \Rightarrow c$ (figure 2).

The indirect link $a \Rightarrow c$ is a *higher-level* link: it does not belong to the original ("plain") structure, but is derivable from it as a collective feature characterizing the structure's integrity. In other words, the structure now has two tiers, one of direct links and another of indirect links. Direct and indirect links are qualitatively different. For instance, direct links can represent the existing physical connections between material things; indirect links then will represent the possible virtual connections. That is, we cannot identify direct and indirect links in a kind of "closure" (yet another mathematical term) for the original structure.

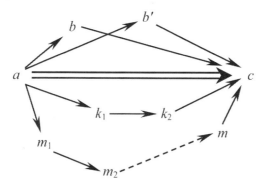

Figure 2. An indirect link.

Now, if two elements of the structure are directly linked, they still can be linked via a number of other elements, and hence there can be an indirect link between them. The combination of both direct and indirect link between two elements of a structure produces a link of the next level, their *connection* in that structure.

# Hierarchical Integrity

In a given set of inter-element relations, some links $a \to b$ may have *inverse* links $b \to a$. However, regardless of the existence of such links, one can consider *inverted* links $a \leftarrow b$ along with direct links $a \to b$ as a sort of higher-level relation, which, taken together with the inverse link $b \to a$, forms a *non-oriented* link $a - b$ on the next level. This is quite obvious from the common viewpoint: if $a$ is related to $b$ as its prototype, $b$ is necessarily related to $a$ as its sequel, and the two elements are thus interrelated. As with direct links, inverted links can be mediated, and indirect inverted links exist as well.

Yet another kind of induced links in a structure is provided by *collateral* links (figure 3). If two elements $a$ and $b$ are linked to the same element $c$, they are already related to each other as having a common sequel. Conversely, if an element $c$ is a common predecessor for both $a$ and $b$, the elements $a$ and $b$ are interrelated. Such collateral links could also be introduced as a combination of indirect and inverted links. Indeed, inverting link $b \to c$, we can then construct a mediated link $a \to c \to b$, which produces indirect link $a \Rightarrow b$. Similarly, inverting link $a \to c$, we get indirect link $b \Rightarrow a$. The combination of the two thus obtained indirect links gives us a non-oriented link between $a$ and $b$.

Figure 3. Collateral links.

So far, only the hierarchy of induced links has been described. However, considering collective mediators and elements reflexively linked to themselves, we can unfold an induced hierarchy of the structure's elements.[3]

To complete the integrity of the structure, "vertical" (interlevel) relations can be added to the picture. For instance, some elements of the structure can be not linked to any other elements. Such elements could be called *irrelevant*, and they must be distinguished from *isolated* elements that are only linked to themselves. Similarly, there can be isolated substructures containing all their links inside themselves. Structures without irrelevant or

---

[3] One can readily observe the resemblance of such induced structures to "thick" propagators in quantum field theory, assuming summation over all virtual "loop diagrams". The divergences characteristic of quantum field theories will not appear in the hierarchical approach due to essentially nonlocal character of induced links and elements.

isolated elements or substructures are called *irreducible*. However, even irrelevant and isolated elements and substructures still belong to the same structure (link from the elemental to structural level), and hence they become interconnected with a subtler indirect relation, and no uncoupled elements remain.

In the induced structure, the very distinction between elements and links becomes relative. Indeed, since any two elements of the structure are somehow connected, any element can thus become a link between links, so that the links will play the role of the structure's elements. This implies that there is no absolute distinction between elements and links in a structure, and the way the structure is unfolded depends from some circumstances external to the structure. Selecting a number of "primary" elements and links, we derive the rest of the structure as induced by the primary set; for another choice, the structure will unfold differently, remaining the same whole. Such conversions make structures *hierarchical*.

## *Systems*

Since the word "system" has penetrated philosophy, it has been applied to almost any kind of things, often used as an equivalent of the pronoun "it". The Greek word "systema" means simply "composition", parts held together in some order, which is closer to structural approach, and people often speak about systems as soon as any inner differentiation takes place. In science, however, a narrower notion of system has long since been adopted. A system is understood as a mechanism of transforming one structure into another in a regular way. Such *functional* stand is used in this book as well.

In general, a system takes some structure as *input* and produces another structure as *output*. This transformation implies that the system is structured itself, and the output may depend not only on input, but also on the system's *state*, comprising both internal and external factors that do not belong to either input or output channels. However, we usually deduce the system's states from how it transforms various structures. That is, systems refer to the *outer* organization of things, rather than their inner organization (structure).[4] Metaphorically, we can speak about a system's "behavior". This, in

---

[4] The limit case of this approach is a stateless system, merely connecting input to output (a black box). When some details of the inner organization of the system are available, it is usually considered as a combination of smaller black boxes.

particular, means that, instead of considering the internals of a system, we will rather combine systems as external to each other, to come to hierarchical systems and systemic integrity.

In the most general case, we picture a system **S** as a collection of *transitions* from an input structure $S$ into an output structure $R$, depending on a structure $C$ representing the system's state:

$$S \xrightarrow{C} R$$

In this scheme, no "inner" details of the system are explicated, since the input and output structures, as well as the system's state, are mere circumstances of a particular transformation and thus rather belong to the system's *environment*[5]. Quite often, a *functional* notation is used:

$$R = F_C(S),$$

where $F_C$ denotes one of a class of functions defined on some class of input structures with the values in some result structure class. In this notation, the structure becomes more "palpable" due to our ability to name its "building blocks" (functions $F$). It should be clear that such a math-like notation does not imply a particular mathematical model, as long as we keep the idea of transformation as a syncretic act, without considering different kinds of transformations and hence guessing about their inner structure. Here, one can imagine a set-theoretic function as mapping from one set to another, or alternatively, take an operational view to a function as an algorithm of computation. Any other model will do as well.

In engineering, systems are often pictured as "functional blocks":

$$S \longrightarrow \boxed{\mathbf{S}(C)} \longrightarrow R$$

Here, a system is regarded as a machine (processor, "chip"), performing certain operations. A simplified inline version of this scheme will be used in this book: $S \rightarrow \mathbf{S}[C] \rightarrow R$. When the system's state is irrelevant to the topic, we can simply write: $S \rightarrow \mathbf{S} \rightarrow R$; conversely, when we are primarily

---

[5] Since we can define a system only through the changes in its environment it produces, each system essentially depends on the nature of its environment. If two different environments can be related to each other, one can also observe that some systems transform related input structures in their respective environments into related output structures. Such systems will be mathematically described with the same equations. However, they are still different, and identifying them would be a logical error. For instance, the motion of a pendulum and the oscillations of electrical current in a special circuit can be described with the same harmonic oscillator model—however, this does not imply that there are mechanical pendulums in the electric circuit, or capacitors and induction coils in a mechanical pendulum.

interested in the system's state, we can write: $S \to [C] \to R$.

Describing systems with the words like "transition", "transformation", "operation" *etc*, we stress the *dynamic* character of a system, in contrast to essentially *static* structures. A structure merely *exists*, while a system is always a change, *motion*.

When the output of some system becomes input for another system, we speak about *sequential composition* (or *cascading*) of systems:

$$S \to \mathbf{S} \to R = S' \to \mathbf{S'} \to R'$$
$$S \to \mathbf{S} \circ \mathbf{S'} \to R'$$

The second (folded) notation encapsulates the very act of composition, representing it as static, structural link. However, in reality, to make the output of one system the input of another, one needs certain operation of transfer, or connection. For instance, it can be physical wiring, or sending a message using some kind of signal. Even in abstract systems, like logic, one still have to somehow identify the output $R$ with input $S'$, which is explicitly denoted by the link $R = S'$ in the first scheme. Obviously, to connect the two structures, $R$ and $S'$, one needs yet another system:

$$R \to \mathbf{M} \to S',$$

and *direct* composition $\mathbf{S} \circ \mathbf{S'}$ thus becomes *mediated* composition $\mathbf{S} \circ \mathbf{M} \circ \mathbf{S'}$. Further unfolding can easily produce a chain of any length at all.

One can observe the resemblance of system composition to indirect links in structures. All the other ways of revealing structural hierarchy have their analogs on the systemic level as well. Thus, collateral links correspond to *parallel composition* of systems, when transforms $S \to \mathbf{S} \to R$ and $S' \to \mathbf{S'} \to R'$ are combined in a single transform

$$\begin{bmatrix} S \\ S' \end{bmatrix} \to \begin{bmatrix} \mathbf{S} \\ \mathbf{S'} \end{bmatrix} \to \begin{bmatrix} R \\ R' \end{bmatrix}$$

or, in the inline notation, $S+S' \to \mathbf{S+S'} \to R+R'$. This implies that input structures $S$ and $S'$ have been combined in a single structure $S+S'$ with a special kind of system, and the resulting structure $R+R'$ has been split into structures $R$ and $R'$ with yet another "splitter" system. For such special systems, structures $S$ or $S'$ (or $R$ and $R'$) can serve as the system's state:

$$S \xrightarrow{S'} S + S' \text{ or } S' \xrightarrow{S} S + S'$$

Their interchangeability here obviously mimics the nonoriented collateral links in structures.

The analog of inverted link is nothing but the very important systemic notion of *feedback*. When the output of the system becomes its next input, we obtain a special case of sequential composition, the powers of the original system:

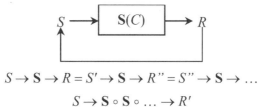

$$S \to S \to R = S' \to S \to R'' = S'' \to S \to \ldots$$
$$S \to S \circ S \circ \ldots \to R'$$

That is, some structure is transformed by the system into another structure, which, in turn, this new structure turns into yet another structure, and so on. Simple transition from one structure to another becomes *a process*. This process can be reinterpreted as a sequence of states of the system. For instance, in a mechanical system, the current state is given with a point $x$ (a vector) in the configuration space $X$, and the process is simply the point's motion in $X$:

$$\ldots \to (x, t) \to (x', t') \to (x'', t'') \to \ldots$$

Here, time moments $t, t', \ldots$ are arbitrary labels used to distinguish one state from the next. One can use any other labeling that will preserve the order of transitions. Systemic motion does not produce time, which must be introduced from a higher level.

Unfolding mediation in the feedback scheme, and using parallel decomposition, we can easily account for the more frequent case of partial feedback, when only a part of the output is made the system's input:

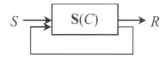

This implies that the system's input and output must be hierarchical structures, unfolded to a number of substructures. The state of the system will also be hierarchical, differently influencing the transition between different substructures.

Different combinations of cascading, parallel connection and feedback produces all kinds of known structured systems. Trying to model the behavior of an elementary (black box) system with some white box, one will discover that different constructions can produce the same outer behavior, exactly like the same structure can be unfolded in different ways. With material systems, even the building blocks can be entirely different in different models. For instance, either metal rod or a brick can support the same construction element; the same computational problem can be solved using very different computers, *etc*. Since we are interested only in what belongs to the system's environment, all the lower-level functioning is considered as *side effect*, and different systems can model each other to that accuracy. However, there is also an analog of the structural conversion: the properties that are considered as side effect in one situation may be essential in another respect.

Various classifications of systems can be found in the literature. Here, I will touch only two of them: open vs. closed systems, active vs. reactive systems. The both oppositions rather refer to the higher level (hierarchies), since they imply hierarchical and developing systems. Thus, an open system is often characterized by exchange of matter and energy with some other systems (possibly included in its environment). However, as long as the system's behavior (transformations of structures in the environment) does not change, such matter/energy flows do not influence the system; if they do, we obviously deal with a higher level system, including the flows of matter and energy into and out of the original system as additional input, output, and augmented system state. The very idea of system's state already implies openness in that sense. It is well known that abstract systems, like theories or activity styles, can also be open, though no energy or matter exchange can be involved in this case. In the hierarchical approach openness we can easily account for that type of openness, since abstract systems can be hierarchical as well as material systems.

Some authors specially stress the distinction of active systems from reactive systems. Reactive systems are said to produce output only when some input is available, while active systems are deemed to be able to initiate some processes in their environment without any input, as well as neglect some input, producing no output.[6] Formally, this means that the system's output is generated from its state, without input; obviously, this is a sign of an incorrect definition of the system, so that some structures have been improperly excluded from its environment and included in the system's state. Conversely, input without output means that some output structures have been erroneously included in the system's state rather than its environment.

---

[6] This distinction is very important in the Argentine neurobiological school, where a psyche's ability of arbitrary action is denoted with a special term (*semovience*). Also, Gordon Allport speaks about the *proactive* character of human behavior, in contrast to mere reactivity of animal behavior.

Such mistakes are quite common with hierarchical, developing systems, where the distinction between the system's environment and its state is relative, depending on the level of hierarchy or a stage of development.

## *Hierarchies*

When we speak about hierarchy, we usually mean that a definite thing is simultaneously considered at different *levels*, and the relations between these levels are of a kind other than the relations inside a level. Today, it is much easier to fancy a hierarchical structure, or a hierarchical system, rather than a hierarchy on itself, regardless of its structural or systemic unfolding.[7] However, it is clearly felt that hierarchy is different from mere system or structure, it cannot be reduced to them. That is, speaking of a hierarchical structure, we assume that the structural picture is (explicitly or implicitly) enhanced with another kind of organization, namely, connecting different structures as the levels of hierarchy. Similarly, in hierarchical systems, a special organization is required to keep the systems belonging to distinct levels together as the levels of the same system. The principle that serves as a "glue" to preserve the integrity in a tiered structure or system is hierarchy proper.

One could say that hierarchy is the synthesis of structure and system, both static and dynamic, and the way of its existence is both being and motion. Development is an obvious candidate for that kind of unity. Indeed, a developing something is treated as changing, but remaining the same thing all the way. On the systemic level, motion is external to the system; for hierarchies, it is the transformation of the system itself that is of importance, so that the system plays the role of structure. Conversely, a hierarchy implies a layered structure that governs the transformations of the system, and hence it functions as a system. Structural and systemic aspects become intertwined, mutually reflected in a hierarchy, as it should be in their synthesis.

Associating hierarchy with development, we can immediately conjecture

---

[7] Since any hierarchy can manifest itself only through a variety of hierarchical structures and systems, one does not immediately perceive the hierarchy itself. Different people take different structural and systemic aspects of the hierarchy for its true shape. Such partial descriptions are often poorly correlated and even controversial, and their advocates can be involved in a both passionate and vain argument, with many quite reasonable observations on every side. These theoretical contradictions are merely apparent, only complementing each other within an integrative view.

## HIERARCHICAL APPROACH

that the levels of hierarchy represent the distinct stages of development. This idea basically agrees with the common tendency to treat complex things as built of simpler things, and serving as the building blocks for yet more complex things. However, hierarchical organization does not exactly fit in that trivial construction. Yes, we can say that a material body is built from molecules, the molecules are built from atoms, and atoms are built from elementary particles, and so on. But an atom cannot be reduced to a number of elementary particles, and a body cannot be reduced to its molecules. A higher-level thing always exhibits *collective* behavior that cannot be explained on the lower levels. Moreover, the higher level motion modifies the motion on the lower levels, producing certain constraints and providing boundary conditions. To put it simpler, the higher levels determine which possible modes of lower level motion will actually take place.

Any individual thing, as soon as it is distinguished from the other things, is also related to the rest of the world. The very distinction of two things is already a kind of relation binding them together. When related to different things, any particular thing manifests its different qualities (the different positions of hierarchy). Eventually, getting in touch with all kinds of things, it will reveal every possible unfolding, thus becoming related to the world as a whole.

Distinguishing what belongs to a thing from what is outside it, we observe that the internal hierarchy of the thing is complemented by the hierarchy of its environment. The inner and the outer hierarchies are mutually reflected. In particular, every particular thing is related to itself through its environment, and hence it plays the role of environment for itself and is reflected in itself. Such reflexive interaction with environment is the principal mechanism of development, the process that changes the very idiarchy of the thing.

The fundamental mechanism of development is *reflexivity*, the thing's relatedness to itself.[8] Such a relation always implies other things mediating this relation. Thus, for structures, we distinguished their elements and links as *internal* to the structure with a look from outside to the structure as a whole. An element of the structure becomes related to itself via its relation to the whole structure. Various feedback schemes in implement reflexivity on the systemic level. On the higher level, the *external* systems that mediate

---

[8] Hierarchies grow due to reflexive interaction with their environment; but that scheme can be inverted, and one could say that the thing's environment reflexively interacts with the thing, and hence it must develop as well, in parallel with the thing's development.

feedback directing a portion of the main system output to its input become the parts of the main system; this is an example of a developing system.[9] In general, reflexivity makes the very distinction between the internal and the external relative, which is an important feature of hierarchies.

Hierarchical development occurs when a number of things form a higher-level integrity, which obviously results in the reflection of this integrity in each component, and hence the growth of their inner hierarchies. That is, hierarchical development is of an *active* character, and things do not merely "undergo" or "experience" some evolution, they change their environment, and change themselves due to the reverse influence of their own products.

While the components of the whole can exhibit their own modes of motion, their belonging to a larger integrity restricts the available possibilities. Only those modes of motion are selected, that support the existence of the whole; the motion of different components is correlated and synchronized.

In science, reflexivity (or self-action) is often represented by *nonlinearity*. Thus, in nonlinear systems, the output structure does not depend solely on the input structures—it can be a product of their combination with input and output of any preceding transforms; in hierarchical systems, which have a definite direction of development, the future can also influence the present behavior, and the more hierarchical is the system, the more it "predicts" its own future in its present.

Any act of an object's interaction with the world implies a cycle of alternating phases of action and being acted upon, which can also be considered as the levels of some hierarchy. When a thing acts on some other thing, it undergoes certain changes; the inverse action partially restores the initial condition.[10] Thus the thing keeps being *reproduced* in every such cycle of action/counteraction, but, in general, not exactly as it was, with some changes gradually accumulated. In the simplest case, such reproduction is reduced to conversion of hierarchy, leaving the object the same and merely changing its form, appearance, or its position in the world. This is referred to as *simple reproduction*; it has to do with all kinds of homeostasis and

---

[9] This produces at least two levels: one corresponding to the "pure" functioning and the other including "self-regulation".

[10] In quantum physics, we find various "virtual" processes of particle emission or absorption. In chemistry, one could point to various catalytic cycles. In biology, there are numerous metabolic cycles. One could conjecture that cyclic motion is more fundamental than mere propagation. A peculiar physics derived from such a principle could be fancied as a complement to the traditional physics essentially based on the idea of inertial motion.

adaptation. Simple reproduction always brings systems to a stationary state, provided there is no external perturbation.

More commonly, things change in their reproduction, which is then said to be *augmentative* rather than simple. In the very common case of *extensive reproduction*, or *expansion*, a larger portion of the world becomes involved in the object's environment, while the character of interactions remains generally unchanged. This results in further unfolding the object's hierarchy. The world becomes deeper reflected in the thing, and the thing imprints itself on a wider portion of the world.

True development (*intensive reproduction*) implies a shift of the boundary between the thing and its surroundings, the change in the very notion of "the internal". This means that the object's hierarchy will change through the synthesis of its own hierarchy with the hierarchy of another thing that formerly was a part of the outer world. This "absorption" of outer things should not be confused with mere consumption. Indeed, consumed things cease to exist; they become entirely disassembled, to provide building blocks for some other structure. This is an extensive process, which is rather characteristic of mere expansion. In hierarchical development, several bodies become involved in some higher level activities, retaining much of their original functionality.[11] One could speak about the formation of a collective body.[12]

As the unity of the internal and the external, hierarchy can develop in two complementary ways, either "zooming in" and unfolding itself into a number of relatively separated inner hierarchies, or growing via binding several things in one. These processes of *differentiation* and *integration* can be mediated or inverted, which can produce very distant mutual influences of things in the world. Virtually, every two things become connected, so that the environment of a thing is reflected in that thing and, conversely, the thing becomes entirely represented in its environment. The whole world thus comes to the state of unity, which, however, is essentially hierarchical: it cannot be comprehended as a given entity, or a process—it is a synthesis of the both.

Saying that the levels of hierarchy represent the stages of its history, we

---

[11] In simple reproduction, thing are reproduced on themselves. In development, a thing is reproduced together with its specific environment, contributing not only to its own persistence, but also to the preservation and growth of other things.

[12] Thus, in biology, individual cells in a specific organ remain relatively independent organisms; on the other hand, organic tissues can change their functionality depending on their place in the living body.

assume that any development can be considered as a sequence of distinct phases. However, the very way of distinction depends on the level of detail, and those considering three stages may be as right as those who distinguish twenty. The process of development is hierarchical itself. Each phase of development can be "split" into many smaller phases, and so on without limit. Conversely, minor changes can be merged in larger units, thus providing a grosser scale for the whole process. Such folding can merge phases in different combinations, and the resulting higher level sequences will be different:

$$...A \to B \to C...$$

unfolds into

$$...A_1 \to A_2 \to B_1 \to B_2 \to C_1 \to C_2 ...$$

which folds to

$$...(A_1 A_2 B_1) \to (B_2 C_1) \to C_2 ...$$

or

$$...A_1 \to (A_2 B_1) \to (B_2 C_1 C_2)...$$

or

$$...A_1 \to (A_2 B_1 B_2 C_1) \to C_2 ...$$

etc.

This is a special case of *conversion* of hierarchies, which makes them exhibit quite different hierarchical structures and systems (the different *positions* of hierarchy), remaining the same integrity. Each of the possible positions corresponds to a possible route of development.

Like any hierarchy, development manifests itself as a number of hierarchical structures, with the levels of hierarchy representing the stages of development. However, because of convertibility, the same hierarchy can manifest itself as different hierarchical structures. This means that, since there are many ways for a thing to interact with the world, development may follow different routes, and different positions of hierarchy indicate the possible directions of its development. This distinguishes the hierarchical approach from other philosophies of development, which either assume a rigid sequence of stages or picture development as a series of random changes. In reality, development is never random, but it may proceed through different stages in different circumstances.

Growth of hierarchies provides the basis of understanding *time*. A cycle of a hierarchy's reproduction provides a natural *time unit*, associated with this particular path of development. Thus defined time must obviously be hierarchical, since every cycle of reproduction looks differently at different levels of hierarchy. There is no fixed collection of reproduction cycles to serves as an absolute "clock". Every hierarchy can exhibit quite different hierarchical structures and hence different *time scales*. This hierarchical time differs from the sheer time variable representing time in physics and many other sciences. The latter is rather a structural parameter, referring to a specific hierarchical structure; in general, time is a measure of the level of development, *hierarchical complexity*. This conforms with intuitive idea of time, implying a definite direction from the past to the future, the existence of a finite "now" within each reflection cycle and the difference in "natural" time flow for different classes of things.

Since any development implies fusion of different hierarchies, the idea of development (and hence the idea of time) is inapplicable to whole world. There is nothing "outer" to the world as a whole, and any distinctions can only happen within the same global entity. However, since any portion of the world can reflect its entirety, each such portion can serve as a world to its inside, and a smaller creature living in such a "world" could conceive the existence of other "worlds", and eventually get in touch with them. However, the birth, existence and death of such partial "worlds" do not have to do with the universality of the world in general, which stays the same, beyond space and time, while incorporating all the possible modes of motion.

## Hierarchical Conversion

Although the basic definition of convertibility has already been presented in the preceding sections, I would like to dwell a little longer on that feature, which is a key point to understanding the novelty of the very idea of idiarchy (hierarchy, in the sense of this book).

Abstract opposition of simplicity and complexity was one of the difficult places in many early philosophies. Since complex things were considered to remain complex for all times, and simple things were thought to be always simple, it was utterly impossible to understand how complex things could be constructed of simple things. Some philosophers therefore denied the very existence of complex things, considering them as mere combinations of simple components (reductionism).[13] The opposite approach was to admit no simplicity and deduce the behavior of the parts of anything from the properties of the whole (holism).[14] As soon as we know that complex structures can be folded in simpler structures, and simple forms can be unfolded without end, we get over this superficial opposition, admitting that simplicity or complexity are only meaningful on a certain level of hierarchy.

Convertibility of hierarchies provides a solid base for integrative studies. This feature simply says that, treating something in a specific *respect*, we deal with its specific *aspect*. The same thing can be involved in many activities and look quite differently in different circumstances.

However, positions of a hierarchy are never arbitrary; they always reflect its overall organization. This is important to distinguish *hierarchy* (idiarchy) from mere *hierarchical structure*—the two categories are often confused in the literature. Though hierarchical structures are normally representations of hierarchies (idiarchies), one can formally construct hierarchical structures that have nothing to do with idiarchy, arranging randomly picked things or ideas in a random way. On the contrary, in any position of an idiarchy, its elements and links are produced from the whole in a specific process of unfolding, according to objective law.

An idiarchy is a strongly connected formation, in which any element is connected to any other. However, connections between its elements are qualitatively different, and the very distinction between the elements and links is relative. Convertibility allows reconnecting the elements of a hierarchical structure in any other order (which implies a change in the quality of their connections); it may mean disappearance of some elements and birth of new elements, so that the initial structure turns into something entirely different.

To grasp the idea of convertibility, imagine a crumpled net lying on the floor in a heap. If you pull one of the nodes, it will drag out the nodes

---

[13] Positivism is one of the most popular forms of reductionism. Also, constructivism has gained strength in the end of XX century. Though the development of science has clearly demonstrated the insufficiency of the both, these paradigms retain their popularity among scientists, who do not pay much attention to methodological consistency and are poorly acquainted with modern philosophy.

[14] In physics, one can observe a model of this controversy in the relations between thermodynamics and kinetic theory; on one hand, physicists are sure that all the macroscopic behavior is due to microscopic motion, and on the other hand they need to introduce various effective forces, fields and potentials, to account for the influence of macroscopic situation on microscopic dynamics.

# HIERARCHICAL APPROACH

immediately connected to it, and they will, in their turn, take out the nodes connected to them, and so on. In the end, you will have the net hanging down from the node you hold, with each node at its own height above the floor. You have produced a hierarchical structure. If you start with a different node, the result will be essentially the same, but the nodes will hang at some other distances from the floor, in a different order. Thus, varying the initial (topmost) element of the hierarchy, you produce different hierarchical structures.

Similarly, dragging up a point of a horizontal cord, you obtain a hierarchical structure ordering the points of the cord by their distance from the flat surface:

Dragging up a different point, you obtain a different ordering of the points:

This new hierarchical structure is called another *position* (or another *turn*) of hierarchy. To understand, why the idea of rotation is invoked, consider another example. In the simplest hierarchy, there are two elements and one link between them. The two possible positions of such a trivial hierarchy can be pictured as

$$\begin{array}{ccc} B & & A \\ \uparrow & \text{and} & \Uparrow \\ A & & B \end{array}$$

Note that the link from $A$ to $B$ is of a different kind as compared to the link from $B$ to $A$, which is stressed by the notation. The example of a triadic hierarchy gives even stronger impression of rotation:

Of course, such simple examples do not convey the whole spectrum of hierarchical convertibility. However, they illustrate how a hierarchically organized thing can turn its different aspects to the world, changing as well as remaining the same in the same time. In addition, the above examples of the net, and the rope, demonstrate yet another important feature of refolding: to get to a specific turn of the hierarchy, the original structure must first be *folded* to some neutral state, and then *unfolded*, starting from a single element that *represents* the hierarchy in this particular position (new hierarchical structure). In the discrete case these two operations are not as obvious, but they still have to be involved.

*Unfolding hierarchies*

The logic of unfolding is based on the relativity of distinction between the elements and links. Thus, in the scheme

$$A \to B,$$

the link $\to$ can be considered as an element $M$ mediating the connection of $A$ to $B$:

$$A \to M \to B.$$

As a result, one obtains three levels of hierarchy instead of the initial two. Any link between the neighboring levels can thus be represented by an intermediate level, and the hierarchy will unfold itself again and again. This is an example of qualitative infinity inherent in any hierarchy.

Once again, one must observe that the quality of links between the elements and levels in any hierarchical structure depends on the way of unfolding, and similar structures may represent quite different turns of hierarchy. There are numerous examples in modern mathematics, where the same notion (*e.g.* a set) can be introduced in the contest of very different conceptualizations (like the number theory or the categorial approach), with all the properties preserved, but in a different sense. Sometimes, this difference can become apparent, like in the case of Riemann and Lebesgue integrals, which coincide in the non-singular domain, but can lead to different results for singular integrands.

Despite of its apparent difficulty, hierarchical unfolding is quite common in our everyday life. Thus, when we first meet somebody, we usually pay attention to some particular details of the person's appearance or behavior, and our further acquaintance with that individual proceeds through extension and moderation of this primary impression. Similarly, to develop a large project, we split it into relatively independent stages, which can further be split into even smaller subtasks.

In nature, hierarchical unfolding is often associated with a fluctuation, a violation of symmetry, or "bifurcation" (in the sense of the catastrophe theory). In any case, this is a natural process, co-relating a thing with its environment.

*Folding hierarchies*

The inverse process of folding a hierarchical structure treats every indirect (mediated) link

$$A \to M \to B$$

as a direct link of a different type:

$$A \Rightarrow B.$$

Intuitively, this corresponds to the common figure of reasoning that, if two things are related through some other thing, they *are* related. The focus shifts from the mediation of connection (its mechanism) to the connection itself, since, in many applications, we do not need to know about the details, as soon as we get the overall result.

Folding is a transition from one hierarchical structure to another structure, which is simpler than the original in certain respects. In our everyday life, a typical example of hierarchical folding is provided by *learning*, when a complex action is first performed operation by operation, but it gradually folds into a single operation that does not require conscious control of the intermediate steps.

In principle, a hierarchy can be folded into a single element; commonly, however, the process of folding stops at some level, with following unfolding in another direction. The "neutral" state, to which the hierarchy becomes folded, can therefore be complex enough,[15] and there can be a hierarchy of such neutral states.

*Multidimensional structures*

In a hierarchy (idiarchy), any element, or link, is a hierarchy itself, and it can be unfolded in its own way, regardless of the current position of the parent hierarchy. Thus, the scheme $A \Rightarrow B$ could become something like

$$\begin{pmatrix} S_2 \\ \uparrow \\ S_1 \end{pmatrix} \to \begin{pmatrix} R_2 \\ \uparrow \\ R_1 \end{pmatrix}$$

Since any part of the hierarchy is connected to any other part, schemes like that always imply missing links, which can be restored in different ways. For

---

[15] For instance, a mountain pass is not necessarily the highest point of the mountain ridge; it's only the most elevated point of the route.

instance, one could consider parallel unfolding of each of the primary levels:

$$\begin{pmatrix} S_2 \\ \uparrow \\ S_1 \end{pmatrix} \rightarrow \begin{pmatrix} R_2 \\ \uparrow \\ R_1 \end{pmatrix}$$

Quite often, however, there is no parallel development of different levels. Thus, the hierarchical structure of the lower level (as the result of its unfolding) can be represented by one of the higher level elements; the rest of higher level development is only indirectly related to the lower level structures:

$$S_2 \rightarrow R_2$$
$$\uparrow$$
$$\overbrace{S_1 \rightarrow R_1}$$

There are many directions of unfolding a hierarchy, and the number of dimensions in the resulting hierarchical structure can grow to infinity. Nevertheless, all the possible unfoldings (positions) of a hierarchy are determined by the hierarchy as a whole and, in that sense, they are contained in it. Every individual thing, at every moment, is in infinitely many relations with the rest of the world, in every one of which it is represented by a specific hierarchical structure. In human activity, that infinity is normally handled using the idea of convertibility, applied to the hierarchy of admissible rotations of hierarchy: at any instance, we only see a particular turn (the topmost element), with the rest serving to enrich it with inner complexity.

## Fundamental Principles

This section summarizes the basic principles that have already been mentioned in the previous discussion of the different levels of organization (structures, systems, hierarchies). Of course, this list is exhaustive; there may be other listings stressing the different aspects of the same. The very idea of a complete inventory of relevant categories and principles is incompatible with the hierarchical approach. However, any practical application requires some mental framework, and this summary could be as useful as any other to grasp the general idea of hierarchy as an intrinsic mechanism of any development.

HIERARCHICAL APPROACH

*Holism*

The category "a hierarchy" conveys the idea of a self-contained thing that remains the same in all the possible contexts. Though it may differently exhibit itself in different respects, all such special manifestations are intrinsically interconnected, being determined by the same organizational center, the whole of the thing. While interaction with the environment is necessary to define to form of the thing and its motion, the thing's development is initiated by its inner dynamics, albeit externally regulated and shaped.

*Hierarchical structure*

Each hierarchy manifests a number of distinct levels, with the higher levels dominating over the lower levels in certain sense; this distinction depends on the aspect of hierarchy under consideration. The elements of an upper level may, for instance, represent classes of lower level elements, or some integral characteristics of lower level motion. In any case, the higher levels are "built" on the basis of lower levels, and they cannot exist without them, despite the apparent higher level control over lower level behaviors.

*Hierarchical system*

At any instance, each hierarchy interacts with its environment as a hierarchical system, transforming some hierarchically structured input into hierarchically structured output. This assumes some inner hierarchy of the system, which can be formally represented by the hierarchy of the system's states. Hierarchical systems are impossible without a hierarchy of feedback channels, and systemic motion is hierarchically structured by feedback cycles. The distinction between "inner" and "outer" structures hence becomes relative, typically determined by the characteristic times of the cyclic processes.

*Infinite divisibility*

The relations between any two levels of a hierarchy constitute a specific entity which may be considered as a level of the same hierarchy lying between the two original levels. Therefore, there is no "final" structure in any hierarchy, since one can always find a new level between any two previously discovered. This procedure will be referred to as *unfolding* the hierarchy.

*Foldability*

The collection of intermediate levels between any two levels of hierarchy can be treated as mere mediation of their connection. All the intermediate levels are thus considered as the inner organization of connection irrelevant to the interrelations of the two selected levels. *Folding* that mediation, we observe the two levels as adjacent. In this way, the total number of levels in a hierarchical structure or system can decrease, and we arrive to grosser view, which presents a logical inverse of hierarchical unfolding.

*Convertibility*

Any hierarchy can be folded, and then unfolded in a different way, hence manifesting a hierarchical structure or system quite unlike the original (another *position* of the hierarchy). Therefore, no hierarchical structure or system should be considered as absolute and rigid; the hierarchy is thus comprehended as the unity of all its possible positions. This multi-faceted nature of any hierarchy is referred to as its *convertibility*, and the transition from one hierarchical position to another is called *conversion of hierarchy* (or *rotation*).

*Relativity of subordination*

Because of convertibility, there is no absolute "topmost level" in a hierarchy, though any hierarchical structure or system will certainly have one. Any element of hierarchy can become its topmost element in some hierarchical structure, thus representing the hierarchy as a whole.

*Strong integrity*

Within hierarchy, the distinction between the elements and their relations can only refer to a particular position of hierarchy, and therefore this distinction is relative. In the same way, any functional distinctions (like input and output, inner and outer) are related to a particular hierarchical system, a specific position of hierarchy.

*Self-conformity*

Any component of hierarchy is a hierarchy too, and it may be unfolded in the same way as the whole hierarchy. The very distinction between the part and the whole therefore becomes relative, and any part of hierarchy may be said to contain the whole of it, the part being virtually equivalent to the whole. To put it differently, a hierarchy is reflected in any one of its elements.

*Qualitative infinity*

Hierarchy does not imply any strict ordering of levels; it rather is a multidimensional formation. The number of its dimensions is "infinite", in the same sense as the number of levels. However, every position of hierarchy implies a one-dimensional ordering of levels, and any level of hierarchical structure or system has a definite dimensionality.

# LOGIC

Logic textbooks define logic as a science about the forms of thinking. Such sentences are twice in error: first, logic is not a science, and second, it is not confined to thinking.

Logic is all about how people choose the ways of doing something, and hence it is a part of philosophy. Any piece of human behavior can be either logical or not (or following different logic in different respects); and every activity has its own logic that may be different from the logic of another activity.

However, unlike the other branches of philosophy, logic decides on acceptability of every individual act judging by formal criteria and disregarding any social circumstances or possible consequences; this gives logic a quasi-objective appearance, as if it were independent of people's interests and concerned only with the natural ways of things. This is why logic may look like (and be mistaken for) science.

Of course, nothing prevents one from enumerating the currently known schemes of reasoning, and that would be a regular science analogous to, say, ethnography. Such a study would never tell a universal logical principle from mere cultural fluctuation. For instance, traditional courses of logic enumerate the forms of syllogisms; but they never tell under which conditions these forms are applicable—and in which cases one should better try something else. Why statements are built of notions? Where do the different truth/verity systems come from? How do people select axioms and basic concepts? To answer these and other similar questions, one needs something more general than science; one has to appeal to the fundamental principles of making all kinds of decisions, including decisions about the adequacy of reasoning. Such principles can only belong to the domain of philosophy.

Since thinking can be considered as a kind of activity, the study of its universal forms (its logic) is governed by the same principles as any other logic study. However, thinking is a very special activity, due to its universality: every conscious action is mediated by thought. That is why the forms of thinking reflect the most common regularities in human activity.[16]

Due to the ubiquity of thought, any activity at all can represent certain modes of thinking; however, some activities are more suited for that than the others—thus, theoretical science (and especially mathematics) may seem to reveal the specificity of reason in a clearer manner because of their abstract nature allowing simple schemes to be implemented in a relatively straightforward way. On the other hand, however, this abstractness would induce the illusion of the subject arbitrarily designing the world at his will, without any concern for what is possible and what is not. In fact, any logical forms can only reflect the position of the conscious being in the real world, and the hierarchy of logical forms will always reproduce (in a specific turn) the hierarchy of the world. Moreover, thinking is not pure logic, and it has to be comprehended through complementary reflections provided by many sciences, as well as the arts, or different branches of philosophy.

To start with, one might indicate that logic reveals universality in the forms of any activity. Primarily, there is the logic of a particular activity—and one could develop it as a separate discipline. In fact, by the end of the XX century, the humanity has accumulated enough ideas about various "special" logics: logic of deduction, logic of interrogation, logic of definition, *etc*. Scientology echoes with such terms as "quantum logic", "situational logic", or "temporal logic". In mathematics, numerous models of different logics have been suggested: many-valued, stochastic, fuzzy, categorial and other logical systems.[17] Unfortunately, most scientists are poorly educated in logic; quite often, they cannot even correctly apply the traditional Aristotelian syllogistics, to say nothing about inductive, modal, or other schemes. Vague notions of logic outside science still hinder the development of human spirituality in general.

Though every kind of activity obeys its own logic, in every culture, some ways of action are of a wider applicability than the others, and there are hierarchical relations between logical schemes. In formal reasoning, we speak about "derivation" of schemes, often believing that there are a few

---

[16] Hence the tendency of reducing all logic to the logic of thinking, and even worse, to the logic of a formal discourse.

[17] However, mathematics is a science, and all mathematical models will always remain within the level of analytical reflection, with its specific logic.

"fundamental" schemes allowing us to obtain all the rest. Usually, simple schemes are considered as more fundamental than complicated.

However, there is no absolute ordering of logical schemes, and formal simplicity or complexity does not imply more universality. In principle, any activity may be made a pattern for many other activities, and the same logical scheme may be obtained in different ways. The unity of logic is rather inferred from the unity of the world, and every two logical schemes can be transformed into each other using the appropriate logical means.[18]

Any logical construction is universal, and the scope of its applicability mainly depends on the cultural factors. There are no restrictions on formal manipulations, and one inference cannot be more logical than another—they only correspond to different cultural situations. However, some ways of actions may become preferable in some societies for quite a long time, thus positioning themselves as universal logical schemes. One can be tempted to declare logic the only source of truth and treat any science at all as a special case of logical development. Under some circumstances this view may be productive enough; but this relation can be easily inverted, so that any other activity (and any science in particular) could become a source of logical schemes. Both, logic is implemented in practical activity, and praxis gives birth to logic.

## Levels of Logic

As any hierarchy, logic can reveal different hierarchical structures. Historically, there was much controversy about the preferable structure; in the hierarchical approach, we understand that there is no preference: each particular structure of logic has its own domain of applicability, and no such structure can fully convey the whole.

### *Syncretic, analytical and synthetic logic*

The adequacy and congruity of activities occurring in people's everyday life is the first manifestation of logic. If one acts according to the natural order of things and the current social expectations, this action is often called

---

[18] The variability of axiomatic systems in mathematics provides a typical example.

a "natural", or "logical", consequence of the objective and social situation. Internal life of a person obeys, from this point of view, its own logic; in particular, the typical routes of thought differ from one individual to another. This level of logic, where the forms of activity are not separated from the activity itself, may be called *syncretic*.

On the higher, *analytical* level, the forms of one's activity become imposed on that activity as external regulations, often codified and officially accepted. For a typical example, take the traditional rules of logic studied by math students as an *a priori* basis of any rigor. More examples: the laws of a state, the rules of a game, editorial guidelines for the contributors to a scientific journal *etc*. Since such forms are relatively independent of the respective activity, their modification may be considered as a matter of convention, since the objectively existing limits of such arbitrary variations may be hidden from people's awareness.

The *synthetic* level of logic assumes that both the rules and their justification become conscious. People may intentionally change the rules for a more adequate behavior in the changing environment, so that no logical scheme is considered absolute and applicable in any situation. This is what philosophy should always desire, though it is only in praxis that synthetic logic can exist as such.

The important corollary is that logical study is applicable to any human activity, as soon as its form implies a universal component. However "irrational" people's acts may seem, they can never be completely illogical, and their logic can be revealed at a closer examination. In particular, since regular thinking is a specific activity, it can be described from the logical viewpoint.

### *Metaphysics, dialectics and diathetics*

In the ancient tradition, logical forms were used syncretically, without too much bothering about the correctness of logical figures, but rather aimed at persuading the opponent to think or act in a certain way. Medieval scholastics continued that tradition, combining rigorous deduction with appeals to the scriptures or common beliefs. However, the first steps of science were to seek for a trustable mechanism of producing new knowledge from the already known, and the axiomatic method of Pythagoras and Euclid seemed to provide the solution. In a couple of centuries, this principle became widespread in Europe, and deductive schemes of ancient mathematics were commonly thought of as the universal language of science and philosophy. The revolutionary discovery of Hegel was to consider everything in development, including the very logic of consideration. Hegel was first to describe different dimensions of logic, and different levels of its hierarchy.

Following Hegel, one can distinguish the following three aspects of any logical act.

In the basis of any practical activity one finds general rationality based on the repetition of the activity's structure. Such *rational logic* deals with stationary activities, where some "standard" forms are preserved for a long time. At the syncretic level, rationality appears as *common sense*; on the analytical level, one can find the traditional modes of reasoning enumerated by Aristotle and widely used in science; in philosophy, this way of reasoning is called *metaphysical*. Though much criticized in the XX century, metaphysical philosophy is a necessary stage of any research, and an indispensable aspect of any thought. It is only in its absolutized form that metaphysics becomes too restrictive and leads to biased opinions rather than knowledge and wisdom.

*Dialectical logic* removes metaphysical rigidity demanding that every action should be viewed in a wider context, along with its alternatives. Everything has its opposite, and the opposites are equally valid, so that the actual activity develops in struggle and mutual reflection of the opposites, and their unity can only be achieved in a higher-level activity. An example of syncretic dialectics is provided by the pragmatic attitude to the world. Analytical dialectics has been widely exercised by the ancient and medieval sophists, and this is the highest form of dialectics possible in philosophical idealism. Synthetic dialectical logic was developed in XIX-XX centuries by K. Marx, F. Engels and their followers. This logic, under the name of dialectical materialism, was to oppose the metaphysical philosophy of early materialists, as well as idolization of logic in idealistic philosophy (positivism).

At its highest level, logic becomes aware of the universal reflectivity, when every category implicitly contains all the other categories, and the whole can be reconstructed starting from any arbitrarily selected element. Unlike dialectics, this logic does not lead to an infinite sequence of levels, the higher ones fixing the contradictions of the lower; rather, it is always aware of the whole hierarchy. Any unfolding of this hierarchy into a sequence of levels according to the dialectical schemes is considered as a particular possibility related to many others, and one arrangement of categories is as admissible as another. Still, these arrangements are not arbitrary, and the rules governing them could be called *diathetics* (intentional arrangement in a specific context). Hegel used to call such logic "speculative" since the only application he could find for it was systematic study of philosophical categories. However, the domain of *diathetical logic* is predominantly in praxis, where nothing reasonable can be done without a clear intention.

## *Levels of intelligence*

The presence of logic in human behavior makes it *intelligent*. Following the usual sequence, from syncretism via analysis to synthesis, one can easily distinguish three levels of intelligence, and the corresponding branches of logic.

1. *Insight.* This level forms the natural basis for any logicality, since no logical development is possible before the object and purpose of activity come to awareness. Thus, any formal definition assumes some previously formed conceptions which do not need to be defined at the current consideration level. Also, to start formal deduction, one has to be aware of the intended result, which cannot be obtained in a deductive way. The ability of insight is commonly known as *intuition*, and it is indispensable for every good scientist.

2. *Discourse.* It is this level that is commonly associated with logic, and the major part of logical research refers to discourse, which may resemble quasi-mechanical application of some pre-established rules. Since, in discourse, the forms of activity become completely detached from the material processes underlying them, reasoning may proceed in quite an arbitrary way, producing abstract combinations of any complexity. The "objectified" character of such logic simplifies its study by scientific methods. The ability of linking several actions in a logical chain is known as *intellect*.

3. *Comprehension.* Neither intuition nor discourse is enough for understanding, which requires a delicate balance of intuition and intellect characteristic of true *ingenuity*. The basis for such synthesis is in the

practical application of logical forms, permanent interaction with the world. In other words, comprehending one logical form requires relating it to another, synthesizing them in a higher-level scheme.

It should be noted that some intelligence may be found in animal behavior too. However, this does not make animal behavior identical to human behavior, since the latter assumes more than mere intelligence: not only should the form of activity be universal, but also its contents. This can only be achieved through communication mediated by *signs*.

### *Regularity and truth*

The levels of logic as described above can only exist as components (or aspects) of a whole, and one level does not deny another, despite their being opposites. The dominance of one kind of logic in a particular activity means that the other logical modes will be used in the same time for self-control, not allowing too much abstraction to lead the activity away from reality. Thus, formal reasoning must be grounded on a sound intuition, and dialectical logic implies formal consistency prescribed by the classical laws; within classical logic, one can observe that basic propositional logic would govern any application of predicate calculus, or any higher-order formal system.

The central category of logic might be called *regularity*. It provides the basis for any logical development, the criteria of logicality, as well as the principles of developing logic itself. Regularity in logic plays the same role as perfection in aesthetics, or ideal in ethics. On the other hand, regularity distinguishes logic from aesthetics, which primarily deals with the unique— and from ethics, which considers regularity and irregularity as two sides of the whole.

Regularity may manifest itself in many ways. However, the most popular form of regularity is known under the name of *truth*, which often was claimed to be the only goal of any logical study. Yes, truth is as attractive in science, as *beauty* in the arts; however, just like beauty becomes artistic when it is enhanced to the degree of perfection, truth may be considered "logical" only if it is pursued *in a regular way*. On the opposite, consistent falsehood may lead to a different logic, which may be more appropriate under certain circumstances. The two logics (based on verification and falsification) are mutually reflected, and this is the cause of their parallel usage in science, where the results obtained within either of the two are often considered equally valid *a priori*, without special reduction procedures required at a higher level of logic.

Philosophical epistemology considers such categories as objective, relative and absolute truth. One may also distinguish rightness, correctness, adequacy, *etc*. The relations between all these categories are to be traced in logic. Another task of logic is to describe the existing logical forms and trace their origin in human activity. In general, logic is to reflect the development of logical forms as a part of the general development of the world and human praxis as its part.

Since different regularities are inherent in different levels of hierarchy, logical truth is also hierarchical. It is only in very simple situations that one can act in a straightforward manner, and each goal is achieved with the only possible action. In most cases there are different ways to do the same, and many different interests are intertwined in the same activity. Normally, at every moment, the hierarchy is found in a certain position, and there is some main logical line; however, all the other lines remain on the lower levels of hierarchy, and they can dominate in another conversion.

### *Discreteness and continuity*

Traditionally, logic is identified with discourse, and thus considered as essentially discrete. Normally, people distinguish one thing from another, and act step by step, thus revealing the discrete side of their activity. However, this does not mean that human activity is all discrete, and there is no place for continuity. Indeed, the distinct operations are embedded in a continuous state of action that lasts from the beginning of the action to its end. The action is also a part of some activity, which does not have a definite beginning or end and might be thought of as purely continuous motion, so that all the discreteness would be treated as limited and virtual.

Here, one could recall conversions of a hierarchy, which reveal discrete structures and functionally differentiated systems in a larger whole that cannot be reduced to any of its particular positions. If logic pretends to reproduce the organization of human activity, it must incorporate means powerful enough to embrace its continuous side.

And this takes place indeed.

For instance, every logical scheme reflects both continuity of activity and its divisibility into separate actions. It is discrete since it contains a finite number of logical positions and junctions. However, both logical positions and logical junctions can be unfolded in different ways, which makes them essentially continuous, though the internal continuity of logical positions is different from the external continuity of logical junctions, and there two different aspects of continuity. However, due to reflectivity, logical positions and junctions are interchangeable, and internal continuity can be transformed into external, and vice versa.

The discrete aspect of logic reflects the objectively developed organization of activity. The two kinds of logical continuity correspond to the infinity of the ways that might lead to the present level of development and the infinity of directions of further development. The present is different from the past and the future, but they are never isolated from them either.

Considering the levels of intelligence, one could observe that discrete forms primarily belong to the level of discourse, while the levels of insight and comprehension involve continuity. However, logic is the unity of discreteness and continuity on every level, and it is only the relative dominance of one or another that varies.

*Organization in logic*

The three fundamental levels of organization are structure, system, and hierarchy. Structures reflect distinction of parts in the whole, as well as their interconnection; this is the static picture taking the elements and links of the whole simultaneously. The dynamic aspect of any distinctions is represented by the system, which manifests possible transformations of one structure into another. As structures form within systems, and systems become elements of a structure, we proceed to hierarchy (or, rather, idiarchy), stressing structural and functional stratification reflecting the directedness of any changes (development).

Considering logic as a whole, one can certainly discover its structural, systemic and hierarchical aspects. Due to self-conformity of any hierarchy, every part of logic will manifest specific structures, systems and hierarchies. The logical aspect of any human activity is thus combining its logical structure (the fundamental interdependencies between the different aspects of the activity), its logical system (the way one stage of activity comes after another), and its logical hierarchy (acquired skills and the directions of their development). Different cultures accentuate different kinds of logic, and there may be practical tasks requiring the domination of the structural, systemic or hierarchical view. That is why people often observe only the dominating level of logic and do not notice the related aspects. However, all the three levels of organization must be present for an activity to be successful.

When logic itself grows into an activity, it develops the same three levels, though different kinds of logic manifest them differently. Typically, there are some logical structures (logical forms) related to each other according to a number of rules or procedures (a logical system). The application of logical rules is regulated by social tradition (logical principles), which determines the possible variations of the logical system.

On the other side, any human activity normally presents itself in a definite hierarchical position, necessarily containing all the other aspects in a hidden way, as the lower levels of hierarchy. For each position of hierarchy, the logic of that activity must come in a specific position too; different logical conversions will reveal structural, systemic and hierarchical logic in the narrow sense of the word, as servicing the structural, systemic and hierarchical aspects of activity correspondingly.

## Classical Logic

Classical logic is one of the most developed parts of logic in general, and its numerous aspects are widely discussed in the literature. However, the majority of books on logic are predominantly concerned with its procedural (empirical) aspect of logic, resembling collections of recipes. The origin of logical rules and the structure of classical logic are still poorly comprehended, and this hinders understanding the other levels of logic, since classical logic forms a natural basis for their development, and they can only be defined in relation to classical logic.

### *What is classical?*

Enumeration of the typical schemes of reasoning given by Aristotle and his school is commonly considered as the origin of logic as a special discipline. However, in Aristotle's texts, formal reasoning was never treated separately from the other aspects of being, including both physical nature and the movements of the human soul. This tradition of philosophical logic has never been interrupted in the course of many centuries, and it continues to the present time. The opposite of classical logic, sophistry, tried to reduce reasoning to mere manipulation with abstractions, and this line has got its clear expression in the modern logical positivism, identifying the schemes of reasoning with reasoning itself, formal models of logic with logic, the form of speech with its content.

Still, classical logic does not cover all the scope of philosophical logic, being concerned mainly with its structural aspects abstracted from their development. This relatively static character makes classical logic most useful in everyday life, while it follows the firmly established cultural standards; however, this inherent rigidity may lead to logical problems in more dynamic situations, where no stable norms could be observed; dialectical or diathetical logic is more appropriate in such cases.

In classical logic, all the objects are supposed to never change during the discourse, so that the whole complexity of their relations could be observed "simultaneously". Of course, one does not mean the physical time here, but rather some "logical time", the order of discourse. Classical logic can be used to treat motion, and even development; but this treatment will always be "classical", that is, accentuating static regularities within any process.

## *Branches of classical logic*

As any logic at all, classical logic is applicable to any activity, and not only to formal discourse. However, traditionally, the ideas of classical logic developed in application to analytical reasoning, which significantly influences the terminology, and most examples in classical logic are about the figures of thought as well.

Due to the universal character of classical logic, various applied disciplines treating the logic of any particular occupation can be constructed. However, the universality of logic also means that such special "logics" will be all like one another, with mainly terminological difference, and hence it is enough to consider one particular object area, to get the logical tools for another. The logic of that scheme transfer also contains a static component that can be treated within classical logic.

Analytical reasoning is rather convenient for logical study due to its essentially formalized character. That is why most logical research has been centered on various formal systems expressible in some natural or artificial languages.

Within this "language-oriented" logic, one could distinguish logic of definition (formation of notions), logic of interrogation (problem formulation) and logic of discourse (currently, the most developed part). Depending on objective relations considered, and the detailed structure of the logical schemes involved, propositional logic, predicate logic, modal logic and many other special logics have been historically formed. A few modern models like multi-valued, fuzzy or categorial logic continue that line, remaining entirely within the scope of classical logic, despite their "alternative" look.

## *Logical forms*

Notions (concepts), statements (propositions) and inferences (arguments) make the hierarchy of fundamental logical forms in classical logic. They all are interdependent, and none of them can be reduced to the others.

The level of *notion* represents the activity of distinction, separating one object from another. Notions are not mere labels of things, they imply knowledge about things in their relation to each other, and hence a notion can be considered as a hierarchy of possible statements about the object.

The notion should not be confused with a word of a natural or artificial language; quite often, there are no adequate words, and lengthy explanations and clarifications may be needed. In many cases no verbal explication can be given at all, and one has to learn notions practically, doing something under somebody's guidance.

*Statements* are built of notions; they relate notions to each other, reflecting the objective relations in the world. Therefore, the number of possible statements is unlimited, since the world is inexhaustible, and ever new relations between notions will reflect additional objective regularities. In a statement, notions are connected in definite order, subordinated to the meaning of the statement as a whole. This integral meaning cannot be reduced to the meanings of the notions involved, and even less to a sentence of natural language or a formal construct; whole books may be sometimes needed to convey the meaning of one sentence, and some relations between notions can only be grasped in practical activity.

However, statements are useless on themselves. They merely express ideas in a form, suitable for further production of other statements, in an inference scheme. Every statement has numerous consequences, without which the sentence has no sense; that is how one comes to the idea of the statement as a hierarchy of possible conclusions.

*Inference* is used to produce new statements (conclusions) from a number of other statements (premises) subordinated within a specific inference scheme. Inference schemes represent the most general regularities of the world, including both nature and culture, and they are usually applicable to many special cases. However, this high level of abstraction results in a higher vulnerability of a conclusion, which is most sensitive to minor shifts in the meanings of the notions involved; this implies that the applicability of a scheme must be substantiated for every instance of its usage.

Like statements represent various relations between notions, inferences connect different relations to each other. Since a notion can be considered as a hierarchy of statements, an inference can also be regarded as a kind of unfolded notion.

As with notions and logical statements, conclusions do not need to be entirely verbal; rather, they are universal schemes controlling the succession of conscious actions within a specific activity. As long as the activity (that is, its motive) remains the same, the consistency of activity can be achieved via logical conclusion.

## *Adequacy, truth, correctness*

It is implicitly assumed that the notions may be either *adequate* or *inadequate*, statements may be either *true* or *false*, and conclusions may be either *correct* or *incorrect*. This dichotomy lies in the basis of classical logic. The adequacy of notions, the truth of statements and the correctness of conclusions cannot be established within logic, requiring inquiry into the relations between the object and the subject, the world and its reflection in human activity. Subjectively, for a logician, this looks like the subject's ability to arbitrarily construct notions, ascribe truth values, or make conventions about admissible conclusions; this arbitrariness reflects the social position of a logician, working with the forms of things abstracted from the things themselves. In reality, logic can only be verified by practical activity, and never by mere formal reasoning. Logic is an instrument for generating hypotheses, and it cannot produce "new" truths from the already established.

The dichotomies of the classical logic originate from a special, but very important activity, binary discrimination, or categorization. The very idea of analytical reasoning implies making sharp distinctions, and opposing a particular thing to the rest of the world. Since analysis is a necessary level of every activity, classical logic is universal and ubiquitous; however, since human activity cannot be reduced to analysis, logic in general is wider than classical logic.

## *Fundamental principles*

Logical principles express the most general, universal rules governing the formal aspects of any activity. Traditionally, three logical principles (or logical laws) are commonly discussed in the literature: the law of identity, the law of non-contradiction (also known as the law of excluded middle), and the law of sufficient justification. However, logical "laws" are not as restrictive as the laws of a science, and they do not determine the exact form of activity, which also depends on the specific conditions of that activity lying outside the domain of (classical) logic; that is why it would be better to speak of logical *principles* rather than laws.

## *The principle of identity*

Definiteness is a distinctive feature of classical logic. Every notion or a relation between notions, or mutual dependence of such relations, is to remain the same during the current activity, which is thus made consistent, in the classical sense. In classical logic, any ideas are merely co-existent; they are defined once and forever, never changing their meaning. The same holds for possible relations between ideas. That is, the principle of identity positions classical logic as an essentially structural approach. Obviously, such a static picture cannot be achieved on the semantic level, since the sense of any word or phrase essentially depends on the context. For instance, the same term can be defined by many different formulations, while the notion is defined as the unity of such partial definitions. This circumstance may lead to communication difficulties, since no finite text can convey the universality of a notion in full; and different people may differently restore the whole from the exposed parts. It is only in common experience and co-operation that the identity of a notion, sentence or conclusion can be maintained—as long as people's activities remain relatively uniform, they will be able to rely on classical logic to organize their social behavior. However, when the society is split to antagonistic classes or exclusive estates, the identity of a notion can only be maintained within the same social group.

*The principle of distinction*

In the act of binary discrimination, a person is to decide on whether one of the two available actions should be taken in response to a specific situation; the basic form of such a decision is: "To do, or not to do?" *Threshold behavior* may serve as a typical model: if a certain quality of the objective situation is intensive enough, the appropriate action is to be initiated. Numerous ways of implementing this dependence lead to many models of logic; all such models refer to the same human ability manifesting itself in different environments.

Everybody can recall situations, when the very act of choice influenced the position of the threshold, thus inducing the denial of the decision almost made. In classical logic, such situations are forbidden, and any distinctions are to be preserved intact within the same activity.[19] That is, once the situation has been put in a particular category, it will always be in this category, and no action may lead to the opposite decision; actions implying opposite categorizations of the same situation are called *contradictory*, and the principle of distinction does not allow combining them in the same activity.

*The principle of completeness*

Any human activity actualizes itself in a hierarchy of conscious actions directed to achieving definite goals. Once the goal is chosen, one has to concentrate efforts on making it closer, which requires a clear view of the goal and rejection of the paths that do not lead to it, as demanded by the principles of identity and distinction. However, one also needs some criteria for terminating the action. Thus, one might decide to stop when the goal of the action has been achieved in full. This is only possible in classical logic based on binary discrimination, so that the any goal is thought to be fully achievable, and any person is thought to be able to distinguish the achieved goal from not yet achieved. The principle of completeness demands that every action should be completed before its results are used in another action. This makes classical logic essentially *sequential*, with all the benefits and

---

[19] This means that classical logic is adequate only within a definite class of activities that do not change the overall situation (or at least the relevant aspects of the situation) too much. In other words, classical logic applies to stable societies in the phase of slow evolution

deficiencies of this approach.

In the sphere of analytical reasoning, this principle takes the form of the law of sufficient justification: a notion is considered as well-defined only if the definition is specific enough and consistent with other definitions; a statement is supposed to be true only if it can be derived from other statements that have already been justified; a conclusion is acceptable only if it based on the complete set of premises and does not go beyond the domain of discourse. In the strictest sense, in formal logic, this principle is formulated as the law of excluded middle: any statement is either true or false (and hence its negation is true), and there is nothing in between; this formulation reveals the inherent insufficiency of classical logic.

*Fallacies*

Within classical logic, any violation of its principles is considered as a logical error. This does not necessarily mean that the results obtained in an erroneous way are themselves erroneous; however, logical errors often have a negative effect, since they are apt to replicate in other similar situations and other logical schemes, which may sometimes result in serious damage to people's well-being. That is why it is important to know about possible logical errors (fallacies) and avoid them.

There are different classifications of fallacies depending on the adopted view at classical logic in general; neither of them can be exhaustive, as there are other positions of the hierarchy that require special consideration. Thus, among the commonly considered, one could distinguish fallacies of relevance, of ambiguity, and of presumption. Fallacies of relevance refer to the arguments relying on premises that aren't relevant to the discussion (for instance, irrelevant appeals). Ambiguity arises in an argument when one connotation of a word is implicitly replaced by another. Fallacies of presumption mean using false premises to derive any desirable conclusion (for instance, false dilemmas and circular arguments). All such arguments (or acts) violate the principle of identity.[20] Other fallacies arise from violating the principles of distinction or completeness.

Nobody is perfect, and every person will make logical errors. Any

---

[20] There are other fallacies of the same class. Thus, logical diversion is very popular in mass propaganda, when the attention of people is diverted by a minor issue presented as more important or urgent.

unnoticed error will result in numerous other errors, and false conclusions, up to apparent paradoxes. The only way to stop this error propagation is to treat any formal results as mere hypotheses, rather than "proofs", and never trust them too much until their validity in their application domain has been practically established. This is a very simple idea: if you are planning to do something this does not mean that you have already done it.

It should be noted that not all fallacies are unmediated. Some people may exploit the others' poor experience with logic to persuade them into wrong actions, using intentionally introduced logical errors. This is one more argument for the necessity of mass logical education.

Fallacies are different from mere delusion. When people do not know something well enough, they may assert something wrong about it, but this is not a logical error, despite its ability to propagate through a sequence of syllogisms. Only when a false statement is intentionally used in an argument, a logical error occurs.

Fallacies should not be confused with *logical paradoxes*. The latter do not violate the principles of classical logic, nevertheless arriving to contradictory conclusions. Sometimes, a false paradox may be encountered, with the results being only superficially contradictory, with a hidden logical error behind the contradiction.

Paradoxes arise in the boundary situations, where the applicability of classical logic becomes problematic; one can never resolve a paradox within classical logic, and a paradox may be considered as mechanism of linking different levels of logic.

## Dialectical Logic

There are numerous books on dialectical logic; however, only few of them are concerned with its specifically logical aspects. This was a side effect of class struggle in the ideological domain, when dialectics became a slogan of one party and a curse for the other. It has been forgotten that the origin of dialectics can be traced up to the most ancient times, and that it was advocated within philosophical idealism no less than by materialists. As any logic, dialectics is universal and cannot reflect the interests of specific social layers. As any logic, it can be used to support quite different ideas, and it is only in practical activity that one way of thought may overcome another.

For a few thousand years, the humanity developed within the three socioeconomic formations based on expropriation of the products of one's activity by individuals or social groups not involved in any production processes; this phase of human development was necessary to break the primitive syncretism of the earliest communal cultures, but its analytical nature manifested itself in all-penetrating social discrimination, and class antagonism. Classical logic was well suited to reflect such a social organization, commonly known as *civilization*. Now, when the last formation of this development phase, capitalism, is approaching its end, accents will shift in all kinds of philosophy to a more dynamic approach allowing for drastic changes and revolutionary development. Dialectical logic perfectly matches this demand.

Unfortunately, dialectical logic was mainly developed outside the English-speaking culture, and it may be difficult to translate many of its categories so that their many-faceted meaning would remain intact. Even in classical logic, a notion could hardly be expressed with a single word or phrase; the more so in dialectics. Up to now, dialectical logic is considered by many people brought up in the classical spirit of stability and determinacy as mere play of words, without any practical importance. Dialectics is difficult to grasp by most scientists, whose essentially analytical occupation forms their minds in a rigid professional mould. It is only in crisis situations that the limitations of the traditional modes of thought become evident, demanding new logical principles to complement the static (structural) approach of classical logic.

### *What is dialectics?*

While classical logic stressed the static, structural aspects of reality, dialectics is all about change. Nothing can remain the same in dialectical logic, and there are no clear shapes and rigid boundaries. The adepts of classical logic would call it absolutely illogical—and it is certainly not logical in the classical sense. However, despite its apparently arbitrary and even chaotic look, dialectical logic remains perfectly rational, being controlled by quite definite principles. As the opposite of classical logic, it is as crisp and formal, and the very its arbitrariness is merely an explicit form of the imminent arbitrariness of abstract classical logicality. And, like classical logic, dialectics can be made into scholastics, if no rapport to reality is maintained.

The motion of thought, and the course of any other human activity, must reflect the motion of the world, for the activity to be successful. This means that dialectical logic, like classical logic, is inseparable from ontology, being the same philosophy viewed in a different aspect.

Classical logic is about quiet things that do not considerably change. Dialectics is the logic of fast changing world, with nothing stable and no time for contemplation. Of course, this situation is as abstract as the absolute rigidity of the classical world. In reality, some aspects of every activity can well be described classically, while dialectical approach is required in other respects.

Dialectical logic says that even though things cease to be the same and transform into something quite different, these changes are not random or arbitrary, they obey certain fundamental rules, albeit very unlike those of classical logic. This explains the practical significance of dialectics, its heuristic value.

## *The origin of dialectics*

Traditionally, Heraclites is said to be the farther of dialectics in Europe. However, dialectical elements can be found in practically any teaching of Ancient Greece, and, of course, in Aristotle's lectures. It is much later that dialectical and classical logic became split and even opposed to each other[21]. In the XIX century, the reverse process of synthesizing the two approaches on a common philosophical basis was initiated, but it is still far from being completed.

As any logic at all, dialectical logic is not an arbitrary construction, and its roots should be sought for in the specific modes of human activity. While classical logic originates from binary discrimination and categorization, dialectical logic is an abstraction of change in general. It is complementary to classical logic in the same sense as considering two distinct things is complemented by considering their difference as a manifestation of their unity. This reproduces the typical place of any distinction within a certain activity, so that drawing the difference between two things is only possible on some common basis. Thus dialectics is implicitly present in classical logic, with its dichotomies being just another aspect of contradictions

---

[21] This was much due the medieval (scholastic) understanding of dialectics as the art of pure dispute, irrelevant to any sense. Newborn scientific thought could not accept that kind of abstract argument, trying to oppose it with "rigorous" classical logic.

inherent in higher-level entities.

It was quite natural to express the ideas contrary to the classical approach in the paradoxical form. Zeno's paradoxes have long since become a standard example. However, dialectics is not mere paradoxes; it can be developed in a positive way, like classical logic. In particular, it has its own logical forms and follows definite principles.

## *Logical forms*

In classical logic, we consider notions, statements and inferences as different levels of hierarchy. In dialectical logic, these forms cannot be considered as distinct enough, as notions or arguments can become statements, statements become notions *etc*, within the same activity. Does it mean that there are no logical forms in dialectics? Certainly doesn't.

In any act of change, there are three aspects universally bound to the very idea: first of all, there is something that changes (thesis), something into which it is to change (antithesis), and the way of transforming the former into the latter (synthesis), something that unites the thesis and the antithesis. These are the basic logical forms in dialectical logic.

### *Thesis*

Anything can change, and hence become a thesis. The possibility of distinguishing the thesis as such implies its relative stability, which open a broad way for applying classical logic to describe it. Notions, statements and conclusions are equally possible as the means of formulation (formalization) of the thesis. However any other aspect of activity that serves as its origin or initial state can be called thesis as well, regardless of whether it can be expressed in words at all. Most generally, in the framework of some activity, its thesis is an objective situation that induces the activity.

### *Antithesis*

As the opposite of the thesis, it is as well abstracted from anything else, and as well describable in a classical manner. The antithesis is a specific thing essentially different from the thesis in some respect. The transformation of the thesis into antithesis necessarily looks like a leap, a jump from one side of a crevasse to another, something unexplainable from the classical standpoint. Quite often, the motive of activity serves as the antitheses to its objective circumstances. We observe the direction of activity from thesis to antithesis, and this is reflected in the standard logical move.

*Synthesis*

The important point in any act of dialectical reasoning is that both thesis and antithesis are the states, phases or aspects of the same thing, which hence must be able to manifest itself in the opposite ways recognizable as thesis and antithesis. Otherwise, this is a quite ordinary thing, which can be described classically as long as its relation to thesis and antithesis is not considered. However, in dialectics, the presence of both thesis and antithesis in the synthetic whole is presented as its inherent *contradiction*. That is, to grasp the synthesis, one must first clearly observe the two opposites, thesis and antithesis, to develop them in full as separate entities, which is known as realization of contradiction. After that analytical part is done, one is ready to connect the opposites to each other and bring them to unity. However, such synthesis is not stable, its inner contradiction leads to a new cycle of contradiction development.

## **Fundamental principles**

While the laws of classical logic have been formulated millennia ago, the principles of dialectical logic had not received a clear formulation until the beginning of XIX century, marked by the works of Hegel and Marx. These formal rules are yet too young to become commonly accepted, or even widely known. Different people will express them differently, but all such formulations are principally the same.

*The principle of integrity*

Dialectics cannot rely on the identity of a thing, since each thing can turn into its opposite under certain conditions. There is a more general principle stating that every definite thing is the unity of its opposite aspects, and that it remains the same despite all the transformations. On the other hand, its internal complexity will make it exhibit its opposite sides to the rest of the world, and each thing has to develop all its possible forms in full until it can cease to exit. Sometimes, the presence of opposite aspects in the same thing may take the form of internal struggle, when two opposite tendencies shape the final appearance of the thing, one of them dominating over another. This is why, in Marxist literature, the principle of integrity was usually called the law of the unity and struggle of the opposites.

From the classical viewpoint, the internal complexity of individual things looks like contradiction in the definition, ascribing opposite attributes to the same notion. In other words, the first principle of dialectical logic says that *every thesis is contradictory*. Applied to the classical logical forms, it implies that no notion statement of conclusion can be fully determinable, and hence any construction based on classical logic is essentially *incomplete*. As negation of the identity of any notion, the principle of integrity was sometimes called the law of contradiction, compared with the law of non-contradiction in classical logic. The idea of dialectical contradiction is a core of dialectics as such.

Practically, the principle of integrity demands that every change were based on the properties of the real things, rather than abstract manipulations. To make anything out of something, one has to use that something according to its inherent tendencies (albeit hidden and non-trivial), and never rape the world trying to make things what they cannot be (the ideological position known as voluntarism).

## *The principle of negation*

While the internal definiteness of a thing is determined by the principle of integrity, the succession of the apparent manifestations of the thing is determined by the principle demanding that every next development phase should be a negation of the original state. In other words, every thesis can (and will) transform into its antithesis under appropriate conditions.

The idea of dialectical negation is quite simple: to produce the antithesis, we have to add something to the thesis that was not in it originally, and, conversely, remove something that should not be present in the result. Adding new features can be considered as removing (negating) their absence. However, in dialectical logic, the changes must be small enough, to preserve the thing's integrity, and there is no absolute change in every respect (which is more like the complement operation in classical logic).

The principle of negation is important to prevent dogmatism. It puts stress on a well-know, but often overlooked, fact that every act is only appropriate in a definite context, and there are no absolute laws, truths, or attitudes.

Dialectical negation is different from negation in classical logic. While the latter leads to an entirely different idea, the former leaves the thing the same, only making it *apparently* (or *functionally*) different; it merely shows how the internal opposites of the thing can manifest themselves in the thing's relation to the world. On the other hand, the negation of negation in classical logic restores the original thing; in dialectical logic, negation of negation is opposite not only to the antithesis negated, but also to the original thesis, negated by the primary negation.

The negation of negation was often said to lead to the thing or situation resembling the original that existed before the primary negation. However, such a view is too simplified to be correct. To return to some features of the original thesis, one must negate the antithesis *in the same respect*, which is not always possible; rather, the negation of negation will result in yet another manifestation of the same thing, which will be different from both thesis and antithesis, retaining them both as its history, and resembling them both, in different aspects. The negation of negation is a *synthesis* of the thesis and antithesis. Any circularity does not belong to the level of dialectics, merely outlining the zones of relative stability, where classical logic could be applied.

*The principle of measure*

The fundamental principle that relates the internal complexity of a thing to its apparent motion via a series of negations says that every definite thing has its *measure*, a unique balance of its internal definiteness (*quality*) and possible external manifestations (*quantity*). The category of quality conveys the idea of a thing as it is, as that very thing, and not another. The philosophical category of quantity cannot be reduced to mere numerical value; it also includes any structural aspects, systemic behavior, or other external manifestations of internal complexity; this is how things of the same quality differ from each other.

Everybody knows that most things can be slightly modified without ceasing to be the same things. Such changes, irrelevant to the quality of the thing, are called quantitative. However, the principle of measure states that quantitative changes can reach a threshold, beyond which the quality of the thing would change anyway, producing something quite different from the original. This is the mechanism of dialectical negation.

The other side of the same principle is that the quality of the thing determines when its quantitative changes will put the end to the existence of the thing as such: everything is the cause of its own death.

It should be noted that, since dialectical negation does not entirely annihilate the thing negated, but rather retains it within the negation, qualitative changes do not produce anything from nothing, merely *transforming* the already existing things, but never annihilating them. Change in quality is still *change*, which implies the retention of something that undergoes the change. This something is reflected in the category of *measure*.

While the principle of negation says that each thing has its limits, the principle of measure states that the limits of a thing are determined by itself. This statement is crucial for the methodology of science, demanding that, for every scientific model, its limits of its applicability should be expressible *in terms of that very model*. One does not need to explain how important the idea of measure is in the arts: it is enough to indicate that, for an artist, the feeling of measure is the principal criterion of achieving the desired result. Also, the principle of measure is a cornerstone of any philosophy, since it is concerned with the very ability to express the infinite and universal in finite and partial philosophies.

## Diathetical Logic

While classical logic deals with static and unchangeable things, and dialectics stresses the aspects of motion and mutability, there logically must exist yet another level, retaining a kind of sameness in any change. Obviously, such logic (I will call it *diathetical*) would be well suited for discussing (and planning) development; formally, it can be associated with the idea of hierarchy (idiarchy), just like classical logic may correspond to the structural view, while dialectics is essentially systemic. However, I intentionally avoid using the term "hierarchical logic" to denote the synthetic way of action combining the features of classical and dialectical logic. Rather, the hierarchy (idiarchy) of logic includes all the three levels, with their interrelations.

Hegel called the synthesis of classical logic and dialectics "speculative logic", which does not seem entirely appropriate. Though this name clearly reflects the active character and main purpose of diathetical logic, its relation to human creativity, it misses the point that logic does not belong to the sphere of thought; it is predominantly manifested in practical activity. In other words, logic is not mere speculation; it is the way of making all kinds of things.

## Why diathetics?

In ancient Greek, the word *diathesis* (and its exact Latin equivalent, *dispositio*) meant intentional arrangement, or a state of being arranged for something. In particular, it was applicable to various representations or exhibitions, as well as the states of mind or moods. The name of diathetical logic stresses this idea of being properly arranged for definite purpose. That is, while classical logic provides standard means to treat any kind of problem, and dialectics says that there are no universally applicable tools at all, diathetical logic admits the existence of suitable instruments for every job, but it also indicates that these instruments may differ from one job to another, and there is a problem of choice. According to diathetical logic, people need to find appropriate ways of solution for each problem, individually selecting from available means. The same goal can be achieved in different ways; there is no unique path to anything. However, every kind of work requires specific methods, and it cannot be done in arbitrary way, applying random instruments in a random manner.

In diathetical logic we use certain logical forms and principles, but we are free to invent new logical forms and revise the very type of our reasoning and action. We are never restricted to common rules, as long as we observe the goal and act purposefully.[22] That is, diathetics implies all-penetrating creativity, including its reflexive application to creativity itself.

## Logical forms

In classical logic, we found such fundamental forms as a notion, a statement (proposition) and an inference. The forms of dialectical logic are thesis, antithesis and synthesis. Both classical and dialectical forms are united in the basic forms of diathetical logic: categories, categorial schemes and paradigms.

### Categories

To start with, one could consider a category as a very general (and virtually universally applicable) notion. However, due its universality, a

---

[22] This does not mean that there is no way for arbitrariness at all. For instance, certain situation may require breaking a causal sequence, to get out from the dead end. Logic violation is quite logical in this case.

category can represent any logical form at all, becoming a synonym of "a logical form in general", something to convey the idea of a certain mode of action.

To use categories, people do not necessarily need diathetical logic. Quite often, categories are used in a particular respect, without explicitly stressing their universality—just like any hierarchy can be unfolded in a specific way. For instance, categories delimit various artistic schools; they distinguish one science from another; any philosophy is developed around a central category, thus becoming distinct from a different philosophy.

One can rarely denote a category with a single word. Outside of context, such a designation is meaningless, it has no sense. The same word can represent quite different categories in different contexts, and people often dispute in vain, since they mean different things under the same words. It is only in action that abstract words can become saturated with definiteness, referring to real situations rather than mere mental constructions.

Since there is no human activity outside a social context, categories are never usable on themselves, without any reference to other categories. That is, a category becomes meaningful only in a categorial scheme, which represents the general conditions of activity reflected in that category. Different positions of the categorial scheme provide distinct connotations of the same category, showing it from various aspects.

*Categorial schemes*

Every logical scheme can be treated as a structure, a system, or a hierarchy. In any case it corresponds to certain (analytical) aspect of human activity, and cognition in particular.

Structurally, a logical scheme contains a number of logical positions (a placeholder for a category) linked to each other with logical connectives. The structural aspect of a logical scheme is commonly used for definition. Every logical position is characterized by a unique collection of properties, and any process of categorization (which is the basis of analytic thought) relates an empirically distinguished object to a position in some logical scheme. Conversely, an object can only be defined by its relations to the other objects, which is reflected in an appropriate logical scheme.

As a system, the same logical scheme may, for instance, describe a number of possible inferences. The systemic aspect of a logical scheme implies splitting it into a number of substructures, and any such substructure is considered as producing the rest of the scheme. Such an "inference" is meaningful only within a particular scheme, and the reliability of the inference depends on the current paradigm. Indeed, the same logical structure can be involved in different categorial systems, thus producing different inferences; however, these differences may be irrelevant to the practical situation, which thus makes a number of categorial schemes equivalent. But this is not the complete equivalence of classical logic, since the differences are still retained somewhere deeper in hierarchy. A categorial scheme can be said to generate hypotheses, which have yet to be practically tested.

From the hierarchical viewpoint, a scheme represents the levels and directions of development. The scheme may be represented as a number of interrelated structures or interacting systems, forming higher-order integrity. There may be many levels, and the resulting hierarchical structure or hierarchical system will represent one of the possible paths of development, from simpler to the more complex schemes. In logical hierarchies, higher levels may be considered as more general than the lower levels; that is, the hierarchical view onto a categorial scheme will readily represent the levels of generality. However, development can proceed in different ways, which become logically related only within a definite paradigm. The same whole can be made of different constituents, and different organization can lead to the same overall behavior. However, that "sameness" is determined by something wider than a categorial scheme

*Paradigms*

Like categorial schemes reflect the levels of generality of categories, paradigms refer to the universality of logical schemes. They distinguish a number of "fundamental" schemes, considering all the others as their specific variants, or representations.[23] Different activities may proceed within the same paradigm, or develop their own paradigms.

A paradigm is the basic mechanism of transferring schemes from one activity to another. It makes people prefer some schemes rather than some others, and reuse them from one activity to another.

Paradigms can also be considered as a mechanism of scheme generation. This process obeys its own logic, which does not fit into classical or dialectical level, though, of course, any particular instance of scheme generation implies both classical and dialectical reasoning. Scheme

---

[23] That is, a paradigm could be roughly comprehended as "a category of schemes", a synthesis of a category and a scheme.

generation is different from logical inference, which assumes a pre-fixed logic. Schemes can be empirically found, derived from the other schemes, or simply suggested for some general reasons, and these three ways are intertwined in the development of logic. Scheme derivation can be integrative (constructing a new scheme from a number of other schemes) or differentiating (unfolding a scheme). The mutual reflection of any object and its environment allows transmutation of logical positions and logical junctions, and this is yet another way of scheme production. However, all these possibilities co-exist within the same paradigm, which determines the balance of the available techniques.

Due to reflectivity and convertibility of hierarchies, there is no absolute distinction between categories, schemes and paradigms. A paradigm can be represented by a category[24], and scheme creation can take form of mere inference.

## *Fundamental principles*

Since diathetical logic has not yet received much attention from philosophers, its basic laws cannot yet be formulated in a comprehensive way. However, certain hints can be found in the literature, as many people have implicitly used that logic in their practical work. Collecting scattered elements from different sources, one could suggest an integrative view.

### *The principle of objectivity*

By objectivity, diathetical logic understands coordination of all the possible treatments of a thing by that very thing. This simply means that people have to account for the objective aspects of the situation in their activity, be adequate and consistent, to achieve anything certain.[25] This applies both to the ways of distinguishing objects in the unity of the world as well as the ways of manipulating with the objects, and the possible ideas about them. If we do something, it necessarily reflects some aspects of how

---

[24] For instance, the "systemic" approach unfolds from a paradigm represented by the category "system", but a paradigm can never be reduced to a single category or categorial scheme.

[25] In discourse, this means that purposeful behavior implies speaking about something certain, rather than merely chatting about nothing. This does not imply that a friendly chat is of no value—rather, in this case, we do not have any discourse at all, and this activity is essentially non-verbal, so that its logic should be rather sought for in actions and attitudes.

the world is organized, regardless of the origin of this organization; natural things are as objective as products of human activity, and social processes are as objective as physical motion or life. The principle of objectivity helps to tell one activity from another, understanding it as the unity of all its structural, systemic or hierarchical aspects. This principle demands to subordinate one's creativity to people's needs, and never waste time and effort on meaningless things. It does not restrict human fantasy, it only direct it to general advantage. Moreover, one can be sure that every human fantasy does not come from nothing; it reflects something in the real world, though this something is not necessarily explicitly indicated, or even consciously recognized.

Inconsistent and purposeless behavior is not compatible with reason; it brings people down to animals. Conscious people can imitate purposelessness for some reasons (for instance, to loosen the grip of tradition and achieve new logicality); but their behavior remains objective and logical, albeit in a different way. Lack of objectivity is always destructive, and no activity can proceed outside particular culture and specific natural conditions. Understanding human behavior requires reconstruction of the basic traits of the objective situation.

*The principle of reflection*

Though people pick out distinct things from the integrity of the world, these things still belong to the world, being interconnected with all the other things. In logic, this leads to the possibility of describing one thing through another; exploring one area of activity, we get some understanding of many others. If somebody has mastered one kind of activity, he can cope with many similar activities, or invent, by contrast, adequate modes of action in complementary activities. Human culture forms a whole, with each part depending on each other. This allows scheme transfer between very distant areas of activity, which produces the impression of logic as a universal basis of activity, though the situation is rather reverse, and it is the all-comprising interrelation of activities that gives birth to general logical schemes.

Hegel spoke of reflective categories that can only be defined through each other, one implying the other. However, the realm of reflection in logic is much wider, as any category is necessarily related to any other. Any logical scheme can be applied to any activity. This does not mean that such arbitrariness will take place in real life. Developing cultures select their own sets of preferences, and scheme transfer itself obeys certain logic. However, if something is not logical in one culture, it may well become quite logical in another.

Reflection in logic is related to the self-conformity of hierarchies. Every logical category, or a logical scheme, represents the whole of logic. Where at least some kind of logicality has developed, all the other kinds are implicitly present as well, and they move higher in hierarchy in its different position.

*The principle of concreteness*

In logic, the general direction of development is from empirical observations to abstract forms, and then to a variety of their practical implementation. Nothing is completely definable within logic, and logical categories originally are mere representations of intuitively felt commonality of things, as well as the human ways of operating with them. It is much later, that such empirical categories become abstract ideas applicable to a wide range of activities and hence irreducible to neither of them. On this stage, one is tempted to admit the primary role of logic in human activity, and forget about its true source, the objective necessity. Logical laws seem to be given us *a priori*, and one's behavior seems to take them as eternal absolute forms. However, abstract principles are utterly inapplicable to real problems, and one has to adapt general ideas to practical needs before they can become regulators of activity. Such practical interpretations manifest different sides of the same idea; however, on the lower levels of logic, they often seem contradictory and incompatible, producing much controversy and public debate.

The principle of concreteness demands that every abstraction should be complemented by a wide range of interpretations, to unfold its real power.[26] Individual acts originally introduced in human activity in a syncretic way, following the objective logic, will necessarily become reproduced as subjective demand, a consequence of one's world vision and convictions. Being abstracted from reality, any idea must return to it as its unifying principle and the common core of superficially different acts.[27]

In other words, consistent logic must grow into practical work. Until

---

[26] This principle was widely speculated upon in Marxism under the name of "stepping up from the abstract to the concrete".

[27] Mere illustrations are not enough; they do not add any concreteness to an abstract idea. However this was the common style of philosophical literature in XIX-XX centuries, and even Hegel, the discoverer of logical development, failed to avoid the temptation of such pseudo-obvious referential explanations.

something is really changed in the world, demonstrating the truth of abstract derivation, logical reasoning is essentially incomplete.

## Formal Logic

The development of the world is characterized by directedness, which produces an ordered sequence of material forms and forms of reflection. Since human society is part of this development, the forms of human activity virtually reproduce the hierarchy of the world. In particular, one's logic reflects one's place in a social group, as well as the position of that group in a larger-scale formation. There are no "truths" equally acceptable to everybody in any situation. However, in every position of hierarchy, all its elements become embedded in a hierarchical structure that determines the objective ways of handling them, and thus the kind of logic appropriate for that type of environment. Correspondingly, individuals as elements of social hierarchy adopt logic assumed by their place in the society.

In particular, every social group is characterized by certain logical peculiarities, reflecting the place of that group in the society. This implies communicating logical schemes from one person to another within the group, so that its members could learn the group logic. However, such learning can rarely be conscious. Typically, it occurs through the process of socialization, along with the development of consciousness itself, which is different from communicating knowledge; rather, this is the matter of correlating individual development with the cultural environment. Still, some pieces of logic can be imposed on individuals as explicit rules, as a part of group knowledge. In such cases we speak of formal logic.

The existence of apparently general logical norms is merely a manifestation of the similarity of cultural positions resulting from the objective structure of human activities. Relatively stable schemes can be historically formed, since cultural development follows a sequence of distinct stages, historical epochs. Such a distinction, of course, depends on the aspect of development concerned, and a few stages can be merged into a single stage on a higher level, as well as any separate stage becomes a succession of shorter historical periods from a closer viewpoint. Nevertheless, in any adopted development scheme, there exists a hierarchy of logical regularities, which can be studied with scientific methods, enumerating, formalizing and interrelating logical norms.

However, logic is not a science, and it cannot be completely comprehended within science. Science constructs models of the world, and logic differs from its formalizations like, for instance, a physical system differs from a physical theory describing it.[28] Any pretence to provide an entirely explicit derivation of some "truth" is bound to fail, because the very idea of truth has nothing to do with science, and people will always have to validate the applicability of scientific "truths" in their practical activity. There can never be entirely explicit reasoning, since the hierarchical nature of human activity makes any explications enthymematic, to an unknown degree. In particular, the major part of any science (including mathematics) is not formal, combining syncretic, analytical and synthetic levels, each with its own logic. Nevertheless, the very presence of logical forms at any level implies that certain aspects of logic can be formalized even outside classical logic, introducing appropriate logical structures, logical systems and logical hierarchies.

## *Classical logic*

Aristotle could be said to be the father of science, and the first science was that of logic. Before Aristotle, philosophers only used some varieties of logic to establish certain "truths" within their philosophical systems, and logical argument was no better for them than poetic metaphors, appeals to common sense, or reference to tradition.[29] As the common logical forms got systematized and analyzed, the ancient syncretism of art, science and philosophy came to its end.[30]

Later Euclid came with his system of mathematics, then Archimedes with his physics and Chrysippus with a new system of logic; all that would be impossible without a solid logical foundation laid by Aristotle.

The formal aspects of classical logic were studied following Aristotle and Chrysippus for many centuries, discovering the forms of notions and

---

[28] And a physical theory is never reducible to its mathematical background.

[29] We do not exactly know whether Pythagoras or his followers used the same logical basis as explicated by Euclid and later scientists. One can admit that the legends of the mathematical achievements of Pythagoreans exaggerated the real situation, and some later findings were ascribed to Pythagoras only by association with the philosophical school.

[30] Of course, this does not mean that the ancient way of though instantly gave way to scientific methodology. It is only in the economic and cultural conditions of the early capitalism that true sciences could appear.

their inner structure, describing the varieties of statements and their combinations, the ideas of consistency and adequacy, the principles of definition, proof, and interrogation. In the XVII century, the logic of induction and investigation came to public attention. However, formal logic had insufficient theoretical basis and did not develop into a separate science, remaining divided between philosophy and various applied fields. In XIX-XX centuries, qualitative logic has significantly degraded, as logical studies have been largely absorbed by mathematics and lost their specificity.[31] Today, the heritage of formal logic is still conserved in some university courses (for law students, future doctors, journalists or editorial workers, *etc*), albeit saturated with modern attitudes, presenting the traditional formal logic as a simplified form of mathematical logic, a sort of popularization for those who are not too good with numbers. Still, formal logic did not disappear; it has only changed its guise.[32]

The ideas of a notion, a statement and inference are among the major achievements of classical logic. In formal logic, these logical forms and relations between them have been given symbolic representation, thus stressing their independence of the possible verbal expressions. In other words, logic deals with some real objects and objective regularities, rather than mere human thoughts and linguistic constructions. The syntax of any natural or artificial language has nothing to do with logic; signs and formulas do not constitute a theory, they merely communicate it. Moreover, logical forms do not necessarily need a verbal expression, since they can manifest themselves directly in human activity.

Formal logic described the possible ways of constructing compound notions and statements using a number of logical junctions (**not**, **or**, **and**, **if…then** *etc*). Despite the universal qualitative difference of these junctions, their meaning depends on the application, and any quantitative models can only refer to a specific position of hierarchy.

Formal logic widely discussed the hierarchy of generality for all logical forms. Notions were divided into individual, general and universal; similar classifications hold for statements and inferences. The form of a simple predicative statement, $S$ **is** $P$, conveys different meanings for different levels

---

[31] Hegel was probably the last to systematically develop logic without reducing it to some science.

[32] One can specialize in logic studying philosophy, mathematics, or computer science. There is no such profession as logician, and the very word becomes associated with logistics rather than logic.

of the generality of $S$ and $P$; the formal consideration of the volume of a notion introduces a hierarchy of notions, and statements of special type are used to express the position of a notion in that hierarchy.

Similarly, combining statements of different generality produces qualitatively different types of inference, and their difference cannot be reduced to merely quantitative difference in the scope of component statements or notions. Inference is not a mechanical transition from one statement to another; it implies qualitative changes and hierarchical ordering.

Yet another important aspect of classical logic is explicated in the formal study of numerous quantifiers, qualifiers and modifiers. Only a small portion of this research has been reflected in the modern mathematical models of logic.

Finally, in classical logic, we come to the idea of a formal system, a regular way of transforming one logical structure into another. Classical logic cannot trace the origin of a formal system, or explain the existence of many logical systems and indicate their interrelations. However, logical systems have great practical significance, and the very presence of some logical system is a clue to mutual understanding of different cultures.

Still, formal logical schemes can never exhaust the content of classical logic. Any science can only approximately represent reality with its models. In particular, truth and falsity, as viewed by a formal model, do not necessarily correspond to logical correctness or incorrectness; the adequacy of a specific way of reasoning (or action) cannot be established within science.

Since science in general tends to oppose people to the world they act in (which is the essence of scientific objectivity), it is no wonder that scientific study of logic often becomes formal, putting forth one structure or another without considering their origin and mutual transformations. The discovered regularities are then viewed as prescriptions, or at least the only possible options. The regulative aspect of formal logic is reflected in the idea of formal truth established once and forever beyond any doubt. Logical science studies the modes of inference that can lead to true statements from true premises; this idea of logical proof has penetrated many spheres of everyday life and common language, and people often mimic formal logical discourse without real need.[33] This could be considered as a kind of logical fallacy,

---

[33] Sometimes, such imitations might even look ridiculous, like many religious writings trying to scientifically derive gods. In philosophy, pseudo-deductive phrasing rather obscures the idea than supports it (*e.g.* Spinoza's *Ethics*).

inadequately appealing to logic where there are more appropriate means of convincing and persuasion.

## *Logic and mathematics*

Any science constructs and investigates formal models of a certain area of human activity—though it may sometimes be difficult to clearly indicate the true scope of the science. Mathematics could be generally defined as a science about the quantitative aspects of human activity. Since the same quantitative aspects are present in quite different activities, mathematics tends to become ubiquitous, easily penetrating any other science.

Formal logic complements mathematics, as it considers universal qualitative aspects of human activity. But quantity and quality always form a unity (measure); a quantitative change can modify a thing's quality, and conversely, qualities imply comparison and hence quantitative characteristics. That is why mathematics and logic have always developed side by side and easily intertwined. However, quality and quantity do not coincide, and formal logic should be distinguished from mathematics, since the two sciences have different scope.[34]

Historically, this distinction was not clearly drawn. Philosophers were often enchanted by mathematics, and tried to give philosophy the same formal look. Leibnitz advocated mathematization of logic yet in XVII century, but it is after the end of XIX century, with the rapid growth of theoretical science[35], that the idea of joining mathematics and logic seized the minds. Some scientists wanted to derive mathematics from formal logic; some others insisted that logic is a part of mathematics. From the logical viewpoint, all such claims fall into a logical fallacy, incorrectly identifying different terms.[36] In general, correspondence (or mutual reflection) does not mean equality, and equality does not mean identity. Adapting mathematical constructions to the needs of logic does not transform logic into mathematics,

---

[34] In diathetical logic, the two sciences could be said to be mutually reflected.

[35] Quantitative theory became very popular under cultural pressure. Industrial engineering demanded accurate knowledge of materials and physical laws. The usefulness of science was judged from the bourgeois standpoint: good science must properly calculate things, to improve technologies and increase profits.

[36] Such exaggerations are quite typical in the era of division of labor, when professions formally separated from each other and opposed to each other; every professional is apt to think that his profession is the most important, most difficult, and deserves better award.

as well as being logical does not reduce mathematics to logic.[37] As usual, other sciences can be formed in the boundary area (*e.g.* metamathematics); one can never draw the line between different scientific disciplines, and attributing some portion of scientific knowledge to formal logic or mathematics often is a matter of personal viewpoint.

Mathematics is a science, and what applies to other sciences applies to mathematics as well. Every science develops from empirical distinctions to theory, which, in its turn, is verified by experiment. Despite its abstract look, mathematics follows the same route. Originally, mathematical objects (a point, a figure, a body, a number) were palpable enough and linked to practical needs. Gradually, new mathematical abstractions were introduced, with rather peculiar and often counter-intuitive properties. However, mathematical objects are rarely invented by mathematicians; usually, they get borrowed from other sciences, which do the preliminary discrimination, outline the range of related phenomena, indicating the desirable quantitative parameters. Mathematicians reformulate the results of other sciences (in particular, other mathematical theories) to fit in the traditional mathematical manner of presentation, which results in more similarity to other mathematical theory and hence more abstraction and more generality. Quite often, however, the rigorous mathematical reformulations of far-from-any-rigor mathematical slang used by other scientists (especially physicists) do not add much to what can be obtained from the loose (and even metaphorical) applied usage of mathematics, while significantly obscuring the principal ideas. To become applicable in practical situations, a mathematical theory needs some informal adaptation, concretization. Such pragmatic testing in special theories plays the role of experiment for mathematics; like in other sciences, experimental results can either support mathematical theory or demand its revision.[38]

Since logical reasoning is a kind of activity, it has its quantitative aspects, and hence can be mathematically modeled. Such mathematical theories (commonly known as mathematical logic) are part of mathematical

---

[37] The same holds for any other science, or non-science, which can be logical with or without a mathematical language (or even without any language at all). In this respect, scientific applications do not much differ from various practical fields, like law, finances, medicine or library maintenance.

[38] Recently, mathematics has got its own experimental ground in computer programming. Computer aided research is becoming a norm of mathematical science, and a number of old mathematical problems could be solved only with the use of computer.

science, with all its advantages and limitations.[39] It can correctly describe certain (quantitative) features of real logic, though, of course, such scientific models of logicality are always narrower that their cultural prototype. On the other hand, logical reasoning has other aspects that belong to the scope of other sciences (for instance, psychology, linguistics, history *etc*), and many scientific models of logic can be constructed. Neither of them is exhaustive, since it is necessary to go beyond science to observe other, non-formal aspects of logicality.

Mathematical logic is no more related to logic than to any other science or application area; the abstract mathematical models can be applied to the quantitative side of the logical aspect of any activity at all. In particular, mathematical logic models certain regularities in mathematical science too. However, true history and methodology of mathematics can hardly be described by mathematical logic. Thus, all varieties of mathematical logic (metaphorically called different logics) are mainly engaged in constructing specific axiomatic systems according to the same formal model: a few inference rules are used to produce all the tautologies of the theory (formulas) starting from a number of tautologies accepted as such without any justification (axioms). Given a collection of elementary statements ("applied" axioms), one can obtain all the true statements substituting elementary statements in the formulas of the theory.[40] Obviously, this is an oversimplified picture of what mathematicians really do. Axiomatic formulation presents a theory as static, as if all the possible knowledge existed once and forever, and the only problem was to formulate the truths contained in the axiomatic definition. In real life, the development of mathematics is not mere quantitative expansion, mechanically increasing the number of theorems proved within the same axiomatic system; there are qualitative changes as well. Euclid would be astonished by what is called geometry today, and Fermat had hardly anticipated the way his last theorem was proved a few centuries after his death. The real logic of a mathematical

---

[39] Traditionally, science names with the prefix "mathematical" refer to the branches of mathematics collecting mathematical theories that have some relevance to the corresponding science: mathematical physics, mathematical psychology, mathematical systems theory... On the contrary, one could also consider psychological mathematics as a branch of psychology (study of the behavioral peculiarities of a mathematician), or logical mathematics (a branch of logic currently known as foundations of mathematics).

[40] The name "calculus" seems more appropriate for such theories: propositional calculus, second order predicate calculus, modal calculus, *etc*.

theory does not coincide with its axiomatic formulation, and real axiomatic systems are far from formal enumeration of terms and inference rules. The definition of the objects of research contains formal and informal procedures, implicit and explicit recursion, inherent contradictions and circular reasoning. Mathematicians have intuitive idea of the object of study before any theory, and formal definitions are initially chosen to contain all the results of interest.

Contrary to the common view, all mathematical logics (Boolean, many-valued, fuzzy, modal, constructivist *etc*) do not differ much from each other, since they develop on the basis of the same methodology of mathematical research and formulated in the same language. A mathematical journal would hardly accept a paper on fuzzy logic containing theorems that are only partially true, or a paper on modal logic with possibly true results. Mathematicians may deny some principles of classical logic—but they will implicitly introduce them in their "non-classical" theories by the way they produce, formulate and communicate their results. Science is a cultural phenomenon, and one cannot break the objective necessity of the scientific method.

Of course the resources of mathematical logic are far from being exhausted, and there always will be new directions of development. One can model non-classical logicality in a mathematical theory, considering time-dependent valuation, hierarchical junctions, variable inference rules *etc*. New application areas will certainly demand more mathematical models. As long as we do not yield to the prejudice of absolute mathematical truth, we can keep on playing with abstractions. However, without reference to the real world, mathematical science can become too formal, more favoring symbolic manipulation than meaningfulness, clarity or practical usability. Similarly, other sciences that largely employ mathematical language sometime get driven away by formal manipulation, replacing scientific research by merely combining symbols, like magic rites aimed to producing scientifically looking theories without too much concern about their actual contents.[41] Though philosophy of science knew sporadic criticisms of primitive

---

[41] Thus, applied physicists well know that there are no true singularities in nature, and any infinitely large or infinitely small values simply mean that the theory has reached the limit of its applicability, and a new theory is required to correctly describe phenomena in the asymptotic region. However, some philosophizing physicists (and popular science writers) take formal singularities for serious, literally speaking about "big bang", "black holes", "catastrophes" *etc*.

rationalization as a model of scientific thinking, the technical efficiency of formal methods seemed so impressive, that, having no as powerful alternative, scientists preferred to never take philosophical speculations for serious, sweeping the stubbornly popping up logical discrepancies under the carpet.

In its extreme form, such denial of scientific methodology arrives to the attempts of reducing science to its language, speaking about the forms of expression rather than the true object of study.[42] This tendency is known as logical positivism.

Mathematical theories should not be confused with any language at all, even with mathematical language. Establishing the objective character of logical forms was a great achievement of classical logic; abandoning this fundamental idea would lead to degradation and decay. One can observe that mathematical language is used differently in different sciences, ranging from metaphor to a formal system; however, no science is reducible to its language, including mathematics. Thus, physicists are known to freely extrapolate mathematical formulas beyond their original meaning, which can pose a lot of problems to mathematicians trying to digest the new mathematical ideas (like path integrals, or asymptotic convergence). But mathematicians themselves always put more meaning to their constructions than any axiomatic formulation can contain. Common language in mathematical theories is not mere simplification; it carries the important function of objectivation, referring to the context of research and its sense.

With the rapid growth of humanitarian knowledge, scientists have to face situations of rapid change and close interdependence of an object and its scientific reflection. This will certainly result in the development of higher level logics, promoting their formalization as well. This may shake the monopoly of mathematical framework in science and the criteria of rigor, as well as significantly change mathematics itself.

### *The basics of diathetical logic*

Since diathetical logic is still in its babyhood, the preliminary outline of its formalism presented here is in no way complete, only introducing the

---

[42] The famous Gödel theorem is an example of how the analysis of a proposition is replaced by a discussion of its linguistic form, explicitly assuming the equivalence of the two activities, which is accepted without any serious justification.

common terms and illustrating the usage of a few typical schemes. The present techniques of formal diathetical reasoning took shape under the influence (and for the needs) of the hierarchical approach; this is not the only possible formulation, and probably not the best.

The important feature of diathetical logic is that any idea develops in several dimensions, since diathetical reasoning is essentially non-sequential. One cannot draw an infinite-dimensioned hierarchy on paper, or even in a 3D medium. All what can be done is to imagine something in three dimensions twisted in ternary and higher-order structures, like complex bioorganic molecules.

As indicated, a category, a categorial scheme and a paradigm are the three fundamental logical forms of diathetical logic. A categorial scheme can be constructed by linking a number of categories to each other like the elements of a structure, and indicating the nature of links. By analogy with classical logic, links between categories are called categorial junctions.[43] In simpler schemes, all junctions are binary; ternary and higher-order junctions can be found in more complex schemes, where they usually represent hierarchies of junctions and can be unfolded in a hierarchical structure with binary junctions only.[44] As follows from the principle of objectivity, categories and the way they are connected in a scheme are not arbitrary, depending on the specific object area under consideration—that is, schemes are validated in practical application rather than by logic. However, the same principle says that any logical operations applied to a valid scheme will produce a valid scheme, provided the limits of applicability are not broken. The danger of such applicability violation is always present since formal representations only partially reflect the hierarchy of a real thing. To avoid logical fallacies, logical operations must be hierarchical, combining both inference and its validation. Thus logic of an object reflects the hierarchy of the object itself. Reasoning (or activity) is called *consistent* in this case.

Consideration the hierarchy of the object together with the hierarchy of its environment leads to hierarchical truth extending the idea of truth inherent in classical logic. In diathetical logic, one deals with different aspects of the same thing, and each position of hierarchy requires its own conceptual framework and, in particular, inference schemes. Direct comparison of such

---

[43] They become logical junctions in a logical scheme, where logical categories are linked to each other.

[44] Such unfolded hierarchical structures are not equivalent to the original scheme, which is the unity of all possible unfolded forms.

partial pictures is impossible; one cannot say which representation is more adequate, they are equally true. Some philosophers praised parallel existence of complementary views as the only valid approach, eclectically combining incompatible descriptions of the same.[45] However, such an approach is unsatisfactory, violating the integrity of the world. In diathetical logic, there is a standard way of synthesizing partial descriptions in a unified scheme: once the complementary pictures are considered as the elements of a hierarchy, there is a new integral view encapsulating all the special aspects and indicating relations between them.

In the simplest schemes, categories are only linked to categories. However, nothing prevents some categories from representing categorial schemes, or even paradigms. Thus categories and junctions themselves become hierarchical; within the enveloping scheme these hierarchies appear in a specific position. The inner organization of categories and junctions is not relevant to the current categorial scheme; it is hidden, undistinguishable, though syncretically contained in the representative category or junction. This kind of encapsulation is known as lift-up (*Aufhebung*).[46]

In the hierarchical approach, any object is treated at different levels, and the relations between these levels are of a kind other than the relations inside a level. The three basic types of organization form a hierarchy too:

1. Structure is an expression of internal (static) complexity. Structure consists of a number of elements, with some relations between them.

2. System refers to external (dynamical) complexity. In general, a system is the way of transforming one structure (input) into another (output), the mechanism of this transformation being defined by the structure of the system's inner state.

3. Hierarchy shows how the external complexity transforms into internal, and vice versa. It orders different structures or systems as the levels of hierarchy, and this order depends from the hierarchy's environment, rather than the properties of its components.

---

[45] Niels Bohr and his principle of complementarity in atomic physics are traditionally mentioned in this context. Further development of physics has demonstrated that the apparent contradictions between the particle and wave models can be lifted up in a unified theory considering new objects that are neither particles nor waves.

[46] The idea of *Aufhebung* was very important in Hegel's philosophy. Unfortunately, it did not attract much attention from logicians during the next two centuries.

# LOGIC

Hierarchical structures and systems are the forms of manifestation of a hierarchy, and development is its native way of existence. Applied to logic, this means that any logical form can be treated as a structure, a system, or a hierarchy; this changes the interpretation of the component forms correspondingly.

*Language outline*

In this book, categories are normally denoted with letters, and junctions are denoted with arrows. The type of a letter, or the form of an arrow, is chosen to express the paradigm used. The structural, systemic and hierarchical interpretation of some typical logical schemes is summarized in the following table.

| Scheme | Structural interpretation | Systemic interpretation | Hierarchical interpretation |
|---|---|---|---|
| $A$ | an element of a structure | a structure | structure, system or hierarchy |
| $\rightarrow$ | link (relation) between the elements; being | transformation of one structure into another; motion | transition from one level to another; development |
| $A \rightarrow B$ | element $A$ is related (linked) to element $B$ | structure $A$ is transformed into structure $B$ | $B$ is the higher level, and $A$ is the lower level |
| $A \rightarrow B \rightarrow C$ | mediated relation; element $A$ is linked to $C$ via $B$ | a process; transformation of $A$ into $B$ and then into $C$ | history; $B$ is an intermediate level, a stage of development |
| $\Rightarrow$ | indirect link | virtual process, implication | subordination of levels |
| $A \Rightarrow B$ | element $A$ is indirectly related to element $B$ | $A$ implies $B$ | $B$ is a necessary stage of development for $A$ |
| — | correlation | systemic interaction | hierarchical interdependence |
| $A - B$ | $A$ is connected to $B$, directly or indirectly | $A$ and $B$ communicate with each other | $A$ and $B$ are interdependent |

Categorial schemes should not be identified with any mathematical constructs; though the graphical representation of a scheme may resemble some mathematical expressions, the meaning of a categorial scheme is beyond mathematics, since the scheme can be interpreted at different levels, always containing the aspect of practical validation. Rather, mathematical theories can be considered as a special case of categorial schemes, their structural aspect.

*Diathetical inference*

In diathetical logic, one logical form can be formally derived from other forms using special logical schemes that can, by analogy with classical logic, be called inference rules. An inference rule can be represented by a categorial scheme; however, every such representation will only refer to a specific position of hierarchy, since inference is a special kind of activity, which is always hierarchical. As any real activity, inference does not need to be sequentially ordered, with one action going after another. Yes, people tend to perform a complex operation subdividing it into a number of simpler steps; however, this is rather a hierarchical unfolding than mere serialization; the beginning of one act does not need to wait for the end of another, and the order is only logical, not temporal. This is especially so with diathetical logic, where inference can start from an apparently arbitrary position and proceed in an arbitrary direction. The final justification of the selected sequences of operations comes from their practical application; formally, this looks like reflexive dependence of inference on the desirable result.[47]

Nevertheless, locally, within a limited range, one can represent inference as a sequence of elementary operations. A list of empirically found formal rules has been compiled within the hierarchical approach:

1. (Reflexivity) If there is a category $K$, there is a junction $\rightarrow$ of this category to itself: $K \rightarrow K$.

2. (Mediation) Any junction $\rightarrow$ can be unfolded into mediated junction $\rightarrow M \rightarrow$, where category $M$ is said to *perform* the original junction $\rightarrow$.

3. (Indirection) Any mediated junction $\rightarrow M \rightarrow$ can be folded in a higher level (indirect) junction $\Rightarrow$.

4. (Inversion) For each scheme $K_1 \rightarrow K_2$, there is an *inverse* scheme $K_2 \Rightarrow K_1$; that is, for each direct junction $\rightarrow$, there is an inverse indirect

---

[47] Even in mathematical logic, inference is not entirely sequential. There are different ways of obtaining the same mathematical result, each implying a specific conceptual context. These derivations are not entirely equivalent; the results are implicitly identified within a certain paradigm, only to achieve the pre-desired closure of the theory. Such inherent logical loops are necessarily present in any proof.

junction $\Rightarrow$, which is normally of a kind different from $\rightarrow$.

5. (Iteration) Each scheme can be multiply repeated: $K_1 \rightarrow K_2$ becomes $... \rightarrow K_1 \rightarrow K_2 \rightarrow K_1 \rightarrow ...$, or $... \rightarrow K_2 \rightarrow K_1 \rightarrow K_2 \rightarrow ...$; the two forms of iteration express reproduction of $K_1$ via $K_2$ and $K_2$ via $K_1$ respectively. The unity of the two cycles is expressed by the non-oriented junction $K_1 — K_2$.

6. (Encapsulation) For any linked categories $K_1 \rightarrow K_2$, there is a category $K$ representing this scheme as a whole (the scheme $K_1 \rightarrow K_2$ is folded into $K$). Indirection and encapsulation are the special cases of lift-up (*Aufhebung*).

7. (Induction) Junctions are reflected in a mediating category: $\rightarrow M \rightarrow$ implies $\rightarrow (S \rightarrow R) \rightarrow$; after mediating the junction in $S \rightarrow R$ with category $C$, the mediated junction $\rightarrow M \rightarrow$ become presented as a contracted form of a triad: $\rightarrow (S \rightarrow C \rightarrow R) \rightarrow$; in the systemic interpretation, the structures $S$, $C$ and $R$ represent the system's input, internal state and output respectively.[48]

8. (Conversion) In accordance with rules 1–5, any scheme $\rightarrow (K_1 \rightarrow K_2) \rightarrow$ can be rewritten as $\rightarrow K_1) \rightarrow (K_2 \rightarrow$, with a proper re-interpretation of the junctions.

Along with these syntactic rules, there are also semantic rules, directing the choice of preferable inferences:

1. (Negation) If there is $A$, there must be some $B$, which is different from $A$ (and hence is *not A*). Further, there is also some $C$ different from $B$; continuing this procedure, one obtains an infinite chain $... \rightarrow A \rightarrow B \rightarrow C \rightarrow ...$ is obtained.[49]

2. (Differentiation) Every $A$ and $B$ are the representatives of the opposite classes: [$A$] and [*not A*], or equally [$B$] and [*not B*]. For every two classes, their typical representatives can be found such that their difference represents the difference of the whole classes.

---

[48] Since this rule can be applied repeatedly, it reflects the *potential* (or *inner*) *infinity* of every object.

[49] This rule came from dialectical logic. The resulting sequence of negations is known in philosophy as *iterative* (or *bad*) *infinity*.

3. (Generalization) Even if *A* is the typical representative of the class [*A*], *A* never coincides with [*A*]; this means that the class [*A*] can be considered as a representative of the class [*not A*].

4. (Integration) Every *A* and *B* can be compared, which implies that they have something in common; this common part of *A* and *B* can be denoted as *C*, representing the join of *A* and *B* (the way, *A* and *B* are taken together).

5. (Lift-up) Every scheme (or a part of a scheme) can be represented by a single category; all the junctions of the original scheme become self-junctions in the folded form (reflexivity).[50]

6. (Duality) If the scheme of an activity reproduces a logical scheme, the activity is said to *implement* the logical scheme; conversely, this activity can also be considered as a *prototype* of the logical scheme. The logical scheme can be treated as the unity of the two aspects of that activity, corresponding to its being a prototype or an implementation.

The numerous special versions of these rules can be formulated, to adapt them to specific applications (methodological variants). Reflectively, such adapted forms can be used as general logical laws too. The formal rules of scheme generation can be arbitrarily combined, producing schemes of any level of complexity. However, these abstract inferences remain meaningless unless there is an implementing activity or a prototype. It is only after practical validation that the newly inferred schemes can be accepted or declined as inconsistent. In real life, this practical validation may take quite a long time; the objective social necessity for a newly discovered scheme may come later, and the scheme remains on the periphery of development until it is rediscovered in its due time.

*Fundamental schemes*

Though categorial schemes can be very complicated,[51] there are a few basic constructions that play the role of logical paradigms. These fundamental schemes are the monad, the dyad, the triad, and the tetrad.

---

[50] This *monad* contains all the possible unfolded forms and hence it could be said to represent *actual infinity*.

[51] For instance, a physical or mathematical theory (or other science) can be considered as an unfolded categorial scheme.

A *monad* is a single category reflected in itself:

The only category in the monad is virtually identical to the only junction. Still, monads are nontrivial, since many other schemes can be obtained from a monad, using appropriate inference rules. The junction as distinguished from the category indicates the context, in which the category $A$ is defined, and hence it refers to the scope of $A$, its place in the world in general.

The structural aspect of the monad corresponds to idea of definiteness, reflecting the persistence of an activity within some temporal and situational range; this constancy (regularity) is the indispensable condition for the very existence of logic. On the systemic level, the monad stresses the inner diversity of any things, which allows considering their different aspects, as well as their mutual transformations within the whole. Hierarchically, the monad reflects the integrity of activity in the course of development; new levels may form, but they will be the levels of the same hierarchy.[52]

In the linear form, a monad can be written as $A - A$, with the non-oriented junction meaning = (equality) on the structural level, $\leftrightarrow$ (mutual transformations) in the systemic interpretation, and the unity of $\rightarrow$ and $\Leftarrow$ (material and ideal reproduction) in the hierarchical sense. According to the general rule, the non-oriented junction (—) implies an infinite cycle of *reproduction*:

$$... \rightarrow A \rightarrow A \rightarrow A \rightarrow ...$$

Structurally, this is simple reproduction; in the systemic reproduction,

$$... \rightarrow A \rightarrow A' \rightarrow A'' \rightarrow ...,$$

the same category $A$ shows its different sides in the cycle of reproduction; in the hierarchical interpretation the same scheme describes development of $A$, with $A'$ and $A''$ representing the higher levels of hierarchy or the new stages of its development.

A *dyad* connects two distinct categories with an appropriate non-oriented junction:

$$A - B$$

---

[52] Applied to science, this paradigm structurally postulates the existence of a definite area of study (the scope of science); on the systemic level, it refers to the normal evolution of science within its scope; as a hierarchy, it allows development of science, extending its scope without making it a different science.

The dyad represents the opposition of two categories, their difference and mutual determination.

There are two complementary positions of the dyad $A - B$:
$$A \to B,$$
which is called the "primary", or "material", position, and
$$A \Rightarrow B,$$
which is known as "secondary", or "ideal", position. The opposition of the primary and secondary junctions is the other side of the opposition of the categories $A$ and $B$. The mutual transition of the opposites in the dyad can be expressed as $A \leftrightarrow B$, or $A \Leftrightarrow B$.[53] In the cyclic representation, the dyad becomes unfolded in an infinite chain
$$\ldots A \to B \Rightarrow A' \to B' \Rightarrow \ldots,$$
with $A'$ and $B'$ being the different forms of $A$ and $B$ respectively. The forms of the dyad can be used to produce schemes of high complexity. In particular, this scheme is used to produce multidimensional dyads:

$$\begin{array}{ccc} A' & \to & B' \\ \Uparrow & & \Uparrow \\ A & \to & B \end{array}$$

*etc.*

The dyad is one of the most popular paradigms, both in common life and in science. The scheme $A \to B$ lies in the foundation of classical logic, representing the universal inner structure of notions, statements and inferences. Stressing the difference of $A$ and $B$, the dyad is a necessary component of any change or subordination. However, this static aspect of the dyad is often exaggerated, leading to the cycle of simple reproduction
$$\ldots A \to B \Rightarrow A \to B \Rightarrow \ldots,$$
with the same categories $A$ and $B$ repeatedly introduced on each stage. This scheme virtually eliminates the very distinction of $A$ and $B$, folding the reproduction cycle into a monad:
$$\ldots \to (B \Rightarrow A) \to \ldots,$$
or
$$\ldots \Rightarrow (A \to B) \Rightarrow \ldots,$$

---

[53] For instance, the primary form of the dyad could express the idea of causation, while the secondary form would mean implication. The mutual forms of the dyad are readily associated with correspondence, or equivalence.

with the two folded forms of the dyad becoming the equality: $A = B$.

The next fundamental scheme, a *triad*, is built of three categories with mutual junctions, or equally, of three dyads:

This is an essentially non-linear scheme, which can be serialized in many ways. For instance, one can use the "cortege" representation $(A, B, C)$ as a synonym of the scheme above.

Each triad has a primary linearization, which is written as $A \to B \to C$; lifting up mediation gives the secondary junction $A \Rightarrow C$; then the full primary position of the triad can be represented by the scheme

The distinction of the primary and secondary positions is relative in logic; in the practical applications, the primary position usually refers to an activity, while the secondary position refers to its reflection.

The primary reflexive (cyclic) representation of the triad is

Here, each category mediates the junction of two other categories. Using the dialectical language, one could observe that category $B$ is a negation of category $A$, and $C$ is a negation of $B$, and hence must reproduce certain aspects of $A$, which is expressed in the primary cycle of triad by "reducing" $C$ to $A$ with a special junction.[54]

In the cyclic form, sequence $B \to C \to A$ produces secondary junction $B \Rightarrow A$, and $C \to A \to B$ produces $C \Rightarrow B$. The secondary linearization of the triad is then given by $C \Rightarrow B \Rightarrow A$, which becomes contracted into the higher-order material junction $C \Rightarrow A$, which, as the negation of negation, is of the same kind as the primary junction $\to$ in $C \to A$. The secondary position of

---

[54] This junction could be called higher-order material junction. It links the triad to itself via the outer world.

# Formal Logic

the triad is therefore pictured as

and the secondary (inverted) cyclic representation is

The primary and secondary cyclic representations of the triad can also be rewritten in the linear format:

$$... A \to B \to C \to A' \to B' \to C' \to ...$$
$$... C \Rightarrow B \Rightarrow A \Rightarrow C' \Rightarrow B' \Rightarrow A' \Rightarrow ...$$

Combining the primary junction $A \to B$ and secondary junction $B \Rightarrow A$, we obtain the dyadic cycle

$$... A \to B \Rightarrow A' \to B' \Rightarrow ...$$

Though $C$ is not present in this cycle, it is still present in the lifted-up form; in other words, $C$ can be said to represent the cycle of reproduction of $A$ and $B$ via each other in the triad. This is expressed in the dyadic unfolding scheme:

$$(A \leftrightarrow B) \to C$$

Similarly, other dyadic positions of triad can be obtained: $(A \Leftrightarrow C) \Rightarrow B$, $A \to (B \leftrightarrow C)$ etc. In the complete dyadic representation, the triad is considered as three interdependent dyads, which gives a higher level of the triad's hierarchy.

The comparison of the primary and secondary positions of the triad helps to understand the difference and mutuality of the primary and secondary junctions. However, like in any hierarchical scheme, categories and junctions are mutually reflected and hence interchangeable; thus, the junctions in $A \to B$, $B \to C$ and $C \to A$ form a primary position of junction triad

$$(\xrightarrow{AB}, \xrightarrow{BC}, \xrightarrow{CA}),$$

the categories $A$, $B$ and $C$ playing the role of the junctions between these junctions. This is the *dual* representation of the original triad.

Formally, a triad can be obtained mediating the junction in a dyad;

however, this will only produce an instance of the triad's isolated position; as shown above, the whole triad contains much more. Still, since a hierarchy can be represented by every its element, the triad can be represented by any of its positions, and the primary sequence $A \to B \to C$ is commonly used to refer to the triad $(A, B, C)$.

The tetrad contains four nodes, and it allows unfolding in numerous dyads and triads. The most common unfolding is given by the scheme $(A, B, C) \to D$, which expresses the logical operation of lift-up (*Aufhebung*), removing the synthesis of $A$, $B$ and $C$ in the triad $(A, B, C)$ and representing the triad as a syncretic unity of its components.

The *tetrad* contains four categories, and it can be schematically depicted as a tetrahedron:

The tetrad is a very complex paradigm, comprising the whole range of possible hierarchical positions. In particular, there are triadic positions like $(A, B, C) \to D$, opposition of dyads like $(A, B) \leftrightarrow (C, D)$, three-level structures like $(A, B) \to C \to D$, and the completely unfolded forms like $A \to B \to C \to D$. Every such position is a separate logical scheme, with its specific meaning. The complete tetrad as the synthesis of all its possible positions is much more difficult to grasp than the monad, dyad or triad—and it is hard to discover in a real activity.[55]

Of course, one might also conceive the schemes of the order higher than four, though such schemes presently seem completely impractical. Most real many-component logical schemes used represent specific unfoldings of some simpler scheme, like the monad, the dyad, or the triad.

*Triads and development*

Traditionally, scientific thought (or at least its formal presentation) is based on classical logic, and the dyadic paradigm. The triad as the simplest extension of the dyad can be used to overcome this dichotomy paradigm and move to the adequate description of development.

---

[55] Still, the individual positions of a tetrad can be found in practical activity, which allows making predictions about the other possible positions

In its primary position, the triad indicates that the junction between any two categories can be considered as a category too:

1. structurally, the link between two elements can be treated as an element of the same structure;

2. in systems, the transform of one structure into another is also structural and the process of transformation is implemented as a specific structure;

3. in hierarchies, the ordering of two levels is represented by some (intermediate) level of the same hierarchy.

This is the simplest model of development through a kind of "interpolation" between the adjacent nodes.

Triads are also convenient for hierarchically unfolding categories, since every category is linked to two other categories, thus mediating their mutual dependence; this leads to distinguishing two opposite aspects in any mediating category; further, the junction between thus introduced "inner" categories can also be mediated, producing a lower level ("inner") triad. That is, the components of the triad can be easily made hierarchical, both potentially and actually infinite. Finally, contraction of the triad to a single category produces a position of a tetrad, with the distinction of the contracted and unfolded levels:

Both in nature and in reflection, triadic formations do not appear as ready-made from the very beginning—they have their own history, reproducing the inner logic of the triad. The typical genesis of a triad ($A$, $B$, $C$) could be outlined as follows:

1. Originally, all the components are syncretically merged together: ($ABC$). This means that the distinctions between $A$, $B$ and $C$ are merely random (unstable) side effect of some lower-order mechanisms;

2. Interaction with other objects supports relatively stable differentiation of $A$ and $C$ as the aspects of the whole; these aspects are yet externally

linked through the outer world $C \to A$. Internally, this distinction looks like a virtual process $A \Rightarrow C$;

3. Interiorization of the external relation $C \to A$ distinguishes yet another component $B$ as an internal material process transforming $A$ to $C$: $A \to B \to C$; this is a lower-order mechanism (implementation) of $A \Rightarrow C$;

4. Reflexive reproduction of the position $A \to B \to C$ in the cycle $\ldots A \to B \to C \to A' \to B' \to C' \to \ldots$ involves a higher-order external junction $C \to A'$. Folding the reproduction cycle in different ways results in the interiorization of this junction, with the formation of virtual processes $C \Rightarrow B$ and $B \Rightarrow A$;

5. The two internal cycles $(A \to B \to C)$ and $(C \Rightarrow B \Rightarrow A)$ become lifted up in a higher-order activity.

This is the standard way of development in the world, including its physical (existential) level, the level of life, and the level of activity and reason. In reflection, this sequence may appear in various inverted forms, since the logic of communicating the results of activity may be different from the logic of activity itself.

# ONTOLOGY OF CONSCIOUSNESS

What is consciousness? The question subsumes many important turns. Here, much more than anywhere else, one cannot be content with sheer definitions, since, with every insight in the nature of consciousness, we change the object of our study, and one would speak not about definition, but rather *determination* (or self-determination) of consciousness, in the broadest sense. Understanding what consciousness is implies understanding the very ability to understand, and one has to also ask how consciousness happens to be reflected in itself, and what it is for.

Primarily, one could apply to the human ability to intuitively distinguish conscious action from physical dynamics, or conditioned behavior—however spurious this distinction may seem in humans. This is the first, immediate determination of consciousness. When this primitive vision is combined with a particular kind of creativity, one would produce either the vivid patterns of art, or the analytical constructions of science, or an ideologically saturated philosophical category. Eventually, these abstractions of consciousness become instantiated in various cultural forms, becoming the *practical* determination of consciousness, its self-reproduction (and development) in human activity.

In a philosophical study, the primary purpose of the ontology of consciousness is to determine the place of consciousness and subjectivity in the hierarchy of the world. This will reveal the roots of consciousness in the non-conscious forms of material motion, and the universal necessity and inevitability of consciousness. Considering consciousness as a specific object, along with any other objects, will stress the unity of the world and overcome a wide-spread tendency of opposing consciousness to the rest of the world and declaring it to be utterly different from all the natural phenomena, supernatural. It is important to both indicate how conscious behavior differs from non-conscious existence, and demonstrate that consciousness is not alien to the world, that it merely continues the line of material development, always requiring a material substrate of a special kind. Also, one will determine, what is the difference between consciousness and subjectivity, and how they are interrelated.

As an immediate consequence of this ontological study, one obtains the universal principles of the inner organization of consciousness, and its general features that do not depend on a particular form of consciousness, or a specific aspect of its manifestation. Numerous hierarchical structures and systems can be found in consciousness, all of them reflecting certain essential moments of its existence.

Finally, one grows to the understanding of the historical nature of consciousness, and considers its development through a sequence of objectively necessary stages and forms, which later become the levels of its internal hierarchy. One has to explain how it happens that some living creatures develop consciousness with time, while the others don't. The historical growth of consciousness, and its unfolding from the most primitive to the higher forms, is to later become a general direction of individual development, ontogeny.

## Ontological Roots of Subjectivity

The first (negative) definition of consciousness is often formulated as "consciousness is what distinguishes a human being from the animal or inanimate body". This definition, however tautological it may seem, conveys a clear idea of a specific feature in conscious beings that makes them essentially different from the rest of the world. Categories like "the spirit", or "an idea", were commonly used as opposites of "matter" and "the body", to express this difference; consciousness was said to belong basically to the realm of the ideal, rather than that of the material. But is this distinction of material and immaterial sides of the world adequate enough?

One historically known solution is provided by *abstract monism*, denying the existence of one of the opposites: either everything is called "matter", in a kind of *vulgar* (naive, intuitive, mechanistic, metaphysical, or natural-scientific) materialism—or, alternatively, everything is claimed to originate from some side of consciousness, like in innumerable variety of idealistic teachings. It should be noted that philosophical idealism is much more diverse than vulgar materialism, since the former is associated with a higher level of abstraction and more distant relation to practical activity. The attempts to substantiate idealism with materialistic elements, and to enrich primitive materialism through somehow accounting for the ideal, may lead to different kinds of *dualism*, which does not relate consciousness to the non-conscious world, but rather admits their independent, parallel existence. Depending on the proportion and arrangement of materialistic and idealistic elements, one could distinguish logical dualism of the Cartesian type, agnosticism, positivism, pragmatism, realism *etc*. This cannot overcome the inherent incompleteness of the two types of abstract monism, since the opposites are combined *in an abstract way* rather than synthesized; they merely coexist as different ideas within the same thought, hindering congruence and consistency.

Abstract monism and dualism are not constructive, in the sense that they try to merely expel the problem of relating consciousness to the non-conscious, so that no further study is possible. In abstraction, the opposites are either identified (everything is conscious, or everything is non-conscious), or they are simply superimposed as entirely independent and unrelated entities. As a result, one cannot speak of the formation or development of consciousness; the ideal is imagined to be eternal and unchangeable, and all the observable diversity of the world is either attributed to the chaotic nature of matter or denied as an illusion, imagination, or a dream.

The only way to bridge the abyss between the conscious and the non-conscious is to admit that consciousness is yet another manifestation of something present in the non-conscious things and processes, different in quality, but not in kind. In other words, consciousness does not emerge from nothing; rather, it forms as a natural continuation of natural development, being yet another level of hierarchy.

But is it possible to preserve the integrity of the world and escape its splitting in two non-intersecting realms, while asserting the qualitative difference of consciousness (or its counterparts in the non-conscious nature) from the material side of the world? Isn't it a kind of dualism too? Yes, it is—if no development is admitted and all the material things are thought to be existing forever in the same form, or a variety of forms, with pre-defined and non-mutable relations between them. No, it isn't—if any distinction is to refer to a specific level of hierarchy, or stage of development, becoming a unity on a higher level.

## The central problem of philosophy

During the several thousand years of analytical culture's development, the humanity knew many different philosophies. In every historical epoch, philosophers discussed the most actual and urgent problems of the time, and the main questions to answer differed from one culture to another, from one historical period to the next. In the class society, the dominant problems of philosophy reflect the current arrangement of social forces at each stage of social development, and philosophical argument reproduces the existing contradictions between the classes. Thus, in XIX-XX century, the question of the priority of matter over mind took the primary importance. Materialism and idealism remain the two major philosophical parties up to now. However, all such local oppositions develop within the same hierarchy, and all individual philosophies remain in the domain of philosophy as such. There is something that distinguishes a philosopher from a political profiteer, and hence there is a common integrative core in all philosophies of all times. It seems like the time for this universal formulation has come.

I conjecture that the principle of the integrity of the world is the cornerstone of philosophy as a specific level of reflection distinct from art and science. The destination of a philosopher is to build a unified picture of the world, and suggest a unified approach to its creative assimilation. Three interdependent aspects can be discovered in this holistic principle:

1. *Uniqueness:* the world is *all*. Nothing can exist "outside" the world, and the very thought of another world puts that "another world" within the world, where the thought has been initiated. There is only one world, and any multiple worlds can only be its parts or aspects.

2. *Universality:* the world is *everything*. The world is diverse, and it is a universe for all its parts, as well as every part of the world plays the role of a universe for its constituents. The world comprises any possible distinction, thus consisting of innumerable partial "sub-worlds", which will be referred to as *things*. The world must shape itself in every possible way, and show all possible manifestations.

3. *Unity:* the world is a whole. Any two things are somehow connected in the world, however different they may seem. Any thing is virtually equivalent to its environment, which complements that thing to the world.

The idea of the integrity of the world in this $3U$ form may seem too general to offer any practical implications. However, it can be further unfolded for any specific demand, and the special philosophy of a particular sort of things can virtually be derived from the integrity of the world. In particular, the integrity of every separate part of the world acting as a world within itself is to obey the $3U$ principle, which gives clues to understanding how a conscious being can create worlds. Yet another immediate consequence of the unity of the world is that every two things in the world have something in common, and phenomena akin to consciousness can be found at any other level. That is, one is certain to encounter analogs of conscious behavior in inanimate or biological systems; consequently, their study helps to comprehend human consciousness, bringing more understanding of the respects, in which inanimate and biological existence is related to conscious, and hence, what in them is of a different kind.

The fundamental principle of the integrity of the world admits that the development of consciousness and subjectivity can be explained, since it is related to other forms of motion and development; there is a way to produce consciousness from non-conscious thing, and higher forms of consciousness form more primitive forms. To illustrate it, let us consider the three aspects of the integrity world that comprise the triad:

$$matter \rightarrow reflection \rightarrow substance.$$

## The world as matter

At any level, the world is comprised of many coexisting things that move and interact according to the natural laws appropriate for that level. There is nothing else in the world, and every phenomenon can only be instantiated in a number of things interacting with each other in a definite way, which constitutes its *material* side. Everything is material in this sense, since everything exists in the same world, and there are no different worlds that would not be a part of the only world embracing them all.

However, for every particular thing, being material does not mean that there is nothing in it except matter. An idea like that is incompatible with the very existence of different things; recognizing only matter, one would not distinguish golden jewelry from sheer bar of gold, or a painting from a dirty rug. Vulgar materialism does not distinguish the properties of things from the things themselves. Such an exaggerated "materialism" can dominate under certain social conditions: thus, a typical bourgeois cannot think of a work of art, or a scientific discovery, otherwise than in terms of money invested in it;

similarly, a hungry person can hardly appreciate fine cookery, until the pains of hunger get discharged.[56]

Speaking of consciousness, a materialistic approach would seek for its material support, the bodily things and their interactions that lead to the phenomena associated with subjectivity. That is, no spirit can exist outside material things, and no explanation of consciousness is possible unless its material substrate is indicated. However, spirit can never be reduced to matter, and one has to find out, in which respect it is different.

*The world as reflection*

The shapes and properties of material things, their arrangement and involvement in other things, their motion and interaction, their development—all those *manifestations* of things are different from their matter, though they would never come without matter. Each thing is characterized by its place in the whole of the world—or in a system of things involved in a common motion—and this is the *ideal* aspect of the thing, as an opposite of material.

Since the world is unique, it cannot communicate with anything else, and any relation of material things is a special case of the world's universal relation to itself, *reflection*.[57] The world is reflected in itself, and it "returns" to itself with every act of interaction, reproducing itself in every instance of development. This reflexivity is as ubiquitous as materiality, and as important for the integrity of the world. It is the "glue" that makes the infinite variety of the world's manifestations into a whole.

The overestimation of reflection is a distinctive feature of philosophical idealism, of either objective or subjective trend. For primitive minds, it looks like magic, that the same material can take so many different shapes, and the same shape can be cast in different materials. Considering reflection in an abstract way as absolutely opposite to matter will necessarily demand inventing somebody, who would impose shapes on raw matter, producing things. Primitive people observed their own ability to produce certain kinds of things, and they fancied that all things in the world (including humans) must be produced by some mysterious being, a god. This is the usual way

---

[56] For the philosophizing bourgeois, the very word "materialism" only refers to metaphysical materialism; the existence of more advanced and more consistent varieties of philosophical materialism (such as dialectical materialism) is simply being hushed up.

[57] This fundamental idea has been explicitly introduced in philosophy by V. I. Lenin in 1908, though it was already inherent in earlier Marxism.

exaggerated abstractions distort the picture of the world.

Consciousness will obviously be related to the ideal aspect of the world, thus being put in the same row as existence, motion, life. That is, consciousness is not material on itself, but it can only exist as a relation between material bodies. The specificity of this relation is yet to be determined, but the very kinship to the other processes and properties in the world is already a solid basis for constructive study.

*The world as substance*

Pursuing the integrity of the world, one must admit that its material and ideal sides cannot exist without each other. The *reality* of each thing is the unity of its materiality and ideality, and the very distinction between the material and the ideal will only refer to a definite position of the hierarchy of the whole, the way it unfolds itself under certain conditions. The world in his reality is *substance*. As substance, the world is both reflected (and reflecting) matter and materialized (and materializing) reflection. This aspect of its unity refers to the self-reproduction of the world. Nothing else is needed to create it, or to trigger its movement and development[58].

Like vulgar materialism resulted from exaggerating the idea of matter, and idealism from overestimation of reflection, the idea of substance as primary to both matter and reflection has historically been made an abstract foundation of a number of philosophies (Spinoza, modern pragmatism, philosophical relativism *etc*). However, isolating the world's substantiality from the material and ideal levels is bound to get lost in unsolvable problems, and it cannot remove the ideological conflict between materialism and idealism. The only true solution will be synthetic, admitting that the material and the ideal are the two sides of the same reality, and they cannot exist without each other. In particular, this means that consistent objective study must consider the ideal component of its object, and virtually its relation to the subject; however, one does not need to introduce consciousness, to describe the non-conscious world, since there are forms of reflection more appropriate to that level.

Every real thing can become a material constituent of a higher level formation—but this would not remove its ideal aspect; the distinction of the material and the ideal is hence *relative*, depending on the level of

---

[58] In particular, there is no need in any deity, or other mythical agent, to explain the appearance of consciousness in the world.

consideration—which, however, does not make them any less opposite. One could observe that what is matter for a higher-level formation is ideal *in a different respect* than what is made of it. There is a hierarchy of both matter and reflection, and any reality is hierarchical as well, the levels of hierarchy reproducing the phases of development. In other words, this hierarchy could be understood as matter becoming reflection, and reflection becoming matter, and it is this mutual penetration that constitutes reality.

To grasp the reality of consciousness and reason, one must understand how the ideal character of consciousness is related to its material implementations. That is, there are certain properties of matter that are indispensable for consciousness formation, and the presence of consciousness is to leave material traces in the world. On this level, the historical forms of consciousness are to be studied, as well as the possible directions of its future development.

## *Levels of reflection*

As indicated, the roots of consciousness are in the ideal side of things, and ideality is hierarchical. That is, we need to find the level of hierarchy, at which consciousness enters the world, and the same level is to be also marked by the appearance of the subject. Presumably, this must be a fundamental distinction, to reproduce the drastic difference of conscious and non-conscious reflection. We know only one as fundamental distinction, that of living creatures and inanimate things. Hence the hierarchy of reflection could be expressed in the triadic scheme, ordering the levels of reflection in accordance with the $3U$ principle of the integrity of the world:

1. *Existence.* This is the most general and fundamental form of reflection, which can be ascribed to anything in the world, including inanimate things. Something must first of all exist, to have any specific features. There may be different kinds of existence, differing by their specific forms of being, motion and development.[59] Such special existences can, in their turn, be hierarchically organized. As existence, a thing syncretically reflects the world, being a part of it; conversely, the thing is syncretically reflected in the world, being virtually identical to its

---

[59] This triad unfolds the hierarchy of reflection by yet another level. As with existence in general, there must be something to be, something to move, and something to develop.

environment. Things and their environment exist via each other.[60] Following one of the possible directions of development, from syncretism to analyticity, one comes to the distinction of inanimate existence and life.

2. *Life*. This is a special kind of existence characterized by the distinction and opposition of an organism and its environment.[61] All the laws of non-organic motion and development apply to living beings as well, but there also are new regularities applicable only on this level. External reflection dominates on the level of life, since an individual organism is essentially a part of the genus, and its relation to the world is mediated by the creatures of the same, or a different kind. While similar indirect relations may be found in inanimate nature as random and optional, they constitute the basis of existence for a living organism, which cannot live without quite definite interactions with other organisms (metabolism). There are different levels of life; some of them are almost indistinguishable from inanimate matter, while some others can overcome the analytical nature of organic reflection, showing the glimpses of universality; in its advanced forms, life becomes aware of the world.

3. *Activity*. This is the most universal kind of life, allowing for subjectivity. The living thing and its environment get re-united on this level due to the formation of an "artificial" environment, culture; however, this unity differs from the syncretism of existence, and the identity of the individual and its environment has to be repeatedly broken and reproduced in a cyclic way. The subject is originally a living creature, but a very special kind of living creature that can be included in the society of other similar beings, reproducing the ways of behavior developed in this society regardless of their immediate physiological significance. This implies a new, *internal* reflection, or *communication*, which serves to transfer the modes of action from one member of the society to another. While similar transfer of behavioral schemes happens in animals only as a transitory feature, communication plays the dominant role in the subject, so that every act is socially oriented, and

---

[60] Like a point particle in a force field.
[61] A living body can even be in antagonism with its environment. Living creatures die in unfavorable conditions; in other cases, they can entirely destroy their environment.

represented in every individual as such, which is called *consciousness*.[62]

The distinction of life from of "coarse" matter (including physical and chemical objects) is an old tradition, as well as the opposition of conscious and non-conscious life. Here, I synthesize the two dyads in a triad, thus conjecturing their common origin in the development of reflection.

The mind, reason, consciousness arise on a certain stage of development, forming a specific level of hierarchy, namely, the social level. However, consciousness must be always associated with some kind of life, and it is certainly related to inanimate existence. Consciousness is not matter, but it cannot exist without a material implementation, which does not need to be unique. The world is hierarchical, and a higher-level formation can be implemented in different combinations of lower level elements, which constitute its material base, while the way of implementation represents its ideal side. This is the germ of consciousness in the inanimate world. Hence, there is no absolute distinction of conscious and non-conscious existence, and one could find a continuum of intermediate levels both between the "physical" existence and life, as well as between conscious and non-conscious life. Still, the level of consciousness is qualitatively different from life and inanimate existence, and it can be represented in any particular biological system only to a certain degree, so that both the form of implementation would restrict the possible manifestations of consciousness, and the participation in conscious acts would influence biological development, leading to the forms that could never be stable without social support.

## *Subjectivity as universal mediation*

Considering the word in general, one comes to the idea of and consciousness as a level of reflection in general. The next step is to look at the world as a collection of individual things, interacting with each other.[63] However, as indicated, a thing cannot exist without a material support, and

---

[62] On this level, living creatures are not merely aware of the world; they also *produce* it, being aware of their products as different from "natural" things. It is through the awareness of the results of one's activity that one becomes aware of one's self.

[63] The words "thing", and "interaction", are used in this context in the widest sense, referring to any kinds of singularity, as well as to any possible way of associating one singularity with another.

there is also an ideal component linking the thing to the rest of the world. Thought both materiality and ideality are relative, there is a quite definite distinction between them on every level of the world's hierarchy, and there can be no motion without something to move, no interaction without something to interact, and no reflection without something to reflect. On the higher levels, any form of life must be associated with a material body (though this body may be comprised of many organisms), and any kind of consciousness must be represented by some cultural body, including one or more organic bodies and a number of inorganic things involved in conscious activity.

On the most general level, the world is reflected (reproduced) in itself, which can be expressed with the scheme $W \leftrightarrow W$; on the analytical level, this recurrence of the world to itself becomes manifested through the interaction of distinct things: $X \leftrightarrow X'$. Relations of every two things in the world are bidirectional; that is, if one thing is related to another, then, conversely, the latter thing is related to the former. Speaking of relation, action, or information, we always mean mutual relatedness, interaction, communication.[64] In the simplest case, the links $X \to X'$ and $X' \to X$ are of the same kind; we say that such links belong to the same level of hierarchy.[65] For example, in classical mechanics, two material bodies act on each other with equal force, but in opposite directions; consider also a chemical bond, symbiosis, a sexual intercourse, a political treaty or a contract. However, it often happens that, for some link $X \to X'$, there is no inverse link of the same level; this does not mean that there is no inverse link at all, since link do not need to be immediate and direct. Things can be linked to each other through another thing, a mediator: $X' \to M \to X$. While both links $X' \to M$ and $M \to X$ may be of the same level, the resulting indirect link $X' \Rightarrow X$ no longer belongs to that level, and it may be as well mediated by a different mediator: $X' \to M' \to X$. The double arrow $\Rightarrow$ here denotes the unity of all the possible paths from $X'$ to $X$, including those with multiple mediations; transition from the collection of mediated links to such a *virtual* link will be referred to as *lift-up* of mediation.[66]

The difference between direct and virtual links is similar to the

---

[64] For simplicity, all such connections of things will be called links.

[65] This case corresponds to mathematical commutativity.

[66] When mediation is lifted up, it does not disappear; virtual links always assume some material implementation, which is merely hidden (folded, encapsulated) on the current level of consideration.

difference between the material and the ideal. And, like the opposition of the material and the ideal is relative, the distinction of direct and virtual links depends on the level of hierarchy.

The inversed logic is also possible. Thus, for a virtual link $X \Rightarrow X'$, there must be some kind of mediation that is characteristic of this very virtual link: $X' \to M \to X$; mediator $M$ is said to *implement* the virtual link $X \Rightarrow X'$. Obviously, implementation does not need to be unique, and the same link can be unfolded differently. However, there exists a special kind of mediator $M(X, X')$, that links $X$ to $X'$; this mediator is said to *represent* the link $\Rightarrow$. Due to the relativity of distinction between direct and virtual links, one can conclude that any link at all is represented by a definite object, and implemented in a number of ways.

Since any link between real things is invertible, mediation of both the direct and inverted links will produce a mediated cycle:

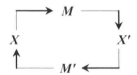

Here, $M$ and $M'$ mediate the links between $X$ and $X'$; similarly, $X$ and $X'$ can be considered as mediating the links between $M$ and $M'$. In principal, nothing makes one choice more preferable than another. Still, in most cases there exist "dedicated" mediators (*signals*), which are more suited to serve as the representatives of interaction; on a higher level, such signals may be represented by material things embodying the lower-level interactions. For example, in physics, one can consider direct interaction of the bodies (*e.g.* Coulomb interaction of atomic electrons); in other situations mediated interaction via a *field* (electromagnetic or other) gives a more adequate picture. However, in quantum physics, fields are always associated with particles (the *carriers* of the corresponding interaction), and the difference between gauge fields (those associated with common interactions) and ordinary "material" fields may become rather vague.

The mediators $M$ and $M'$ may be of the same kind (like electromagnetic field mediating the interaction of charged particles); alternatively, the direct and reverse processes will be mediated by mediators of different kinds (like in catalytic reactions in chemistry, or biological cycles). In the first case, all interactions occur within the same level of hierarchy. The second case also allows such a "planar" consideration, but there is yet another option. Thus, if the mediators $M$ and $M'$ are qualitatively different, we could treat one of them (say, $M$) as belonging to the same level as $X$ and $X'$, while the second mediator would be put "outside" the system $X \to M \to X'$, providing a kind of environment for it:

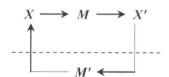

Such a representation implies that $X$ and $X'$ are, in a sense, "closer" to $M$ than to $M'$; for instance, the characteristic times of the corresponding transitions are of the same order for transitions in $X \to M \to X'$, while the "external" mediation via $M'$ takes much longer times. This approach is related to distinguishing the levels of mediation by the degree of their indirectness. However, due to relativity of that distinction, one could expect that separating a system from its environment can be a non-trivial task.

Discussion of the hierarchy of links is not as abstract as it may seem. Thus, formation of associations in animal and human psychology provides a vast area of application for hierarchical approach. Each association forms on the basis of certain material processes, and is directed by external conditioning. However, once association (virtual link) has formed, it becomes represented by certain environmental changes, which support reproduction of particular mediations. The existence of a characteristic mediator for every link is of crucial importance for understanding consciousness. One can conjecture that virtual links have to do with consciousness itself, while the corresponding characteristic mediators are readily associated with conscious beings. However, not any virtual link can be related to consciousness, and one comes to considering the hierarchy of mediators.

Though this hierarchy can unfold in many dimensions, the universality of mediation seems to be the most appropriate for determining the domain of reason. We observe that different types of mediation dominate on the levels of existence, life and activity. Thus, the inanimate world knows only passive mediation: coexistence, intermediate states of motion, correlation *etc*. Mediated interaction is one of the most important cases. In the chain $X \to M \to X'$, the signal $M$ is *emitted* by $X$ and *absorbed* by $X'$; in many cases, $X$ and $X'$ continue to exist as they were, with only their state of motion changing. The mediators are often of the same kind as the things connected, and they behave like particles, waves *etc*. Some inanimate things seem to be unable to interact via certain interactions (for example, strong interaction of leptons is forbidden). However, in general, any inanimate thing can interact with any other thing, and hence mediate interactions of other things. Such mediated connections between inanimate things are *random*, in the sense that they are not necessary for the existence of the things themselves. For instance, an isolated electron will still remain an electron, and a molecule does not need to interact with other molecules to become a molecule of that very sort. Interaction is not needed to support the existence of inanimate things, and often leads to their destruction or transformation into other things.

The animate nature is characterized by active mediation, with the mediator $M$ *consuming* thing $X$ and *producing* thing $X'$. Unlike the interaction of inanimate things, $X$ does not exist any longer after it has been consumed, and $X'$ did not exist before it has been produced by $M$. On this level, $M$ is not merely effectuating the interaction between $X$ and $X'$—now, it is seeking for $X$ to produce $X'$. Moreover, the very existence of a living thing (an *organism*) depends on its ways of consumption and production, and terminated metabolism means death. This is the level of *necessary* mediation. On this level, mediations become essentially asymmetrical: the processes within an organism are often clearly distinct from its interaction with the environment. However, this does not mean that the organism can be defined on itself, isolated from its connectedness to other organisms and the inanimate world. On the contrary, an animal is essentially a representative of its species, and a part of an ecosystem; if a group of animals becomes isolated from their natural habitat for a long time, this will result in either their degradation and death or transformation into another species adapted to the new environment.

Like inanimate things can be joined by mutual interactions into a composite body, organisms tend to cling together forming a higher-level organism. However, such coexistence is much more restrictive, since any of the organisms living together requires a quite definite environment to live. In every particular synergy of different organisms, the members of this communion have to adjust their structure and behavior, to serve the whole. Thus, the organs of the animal body, while remaining relatively independent organisms, are functionally dependent on each other, and evolve to the forms, which cannot live outside the body; compare this with the molecules in a solid body: while slightly changing in the solid body's structure, they can always be separated from it, and still continue to exist. Among other examples of hierarchical organisms, uniting and transforming lower-level organisms are a cell as a system of organelles, an organ as a community of cells, a colony of insects as a collective organism, a biocenosis, a biosphere in general...

The relation of the living organism $M$ to the things $X$ and $X'$ connected through it is relatively rigid, pre-defined, characteristic of the species. It is only in higher animals that more flexible types of behavior become possible, and $X$ is not necessarily consumed to produce $X'$, and different ways of consumption are possible to produce the same thing. In particular, some organisms of the same species can serve as triggers for certain organic processes, not being directly involved in them—in a sense, this resembles catalytic reactions in chemistry. The mechanism of such communication between higher organisms is still based on innate metabolism, with one animal producing a thing that is consumed by another animal, which leads to behavioral changes in the latter; however, this kind of consumption is not immediately related to the basic metabolism that supports the animal's life. In other words, the behavior of complex organisms becomes hierarchical, involving both the level of vital functions and the level of conditional functions supporting the organism's ability to maintain its basic metabolism, and life. In higher animals, the support functions significantly overweigh the basic metabolism; this dominance is a premise of consciousness formation.

It is only on the level of subjectivity that mediation becomes really *universal*, and any two things can be linked through the mediator of a new type, the *subject*. This universality of mediation differs from random mediation on the inanimate level, since it is necessary for the subject, and any subject is bound to bring things together, to remain a subject. However, this also differs from organic mediation, since it is no longer restricted to a specific class of links and extends to the whole world. Such an all-embracing necessity is called *freedom*.

While inanimate mediators link things only in their immediate environment, and living creatures can effectuate only those transformations that are compatible with their physiology, the subject can link anything to anything, with no physical or physiological limitations. Things that do not directly interact, to any appreciable extent, become united in the activity of the subject, thus restoring the unity of the world. For instance, there is no physical reason for the Polar star to influence the movement of a ship—and no physiological reason for the human organism to react on the starry sky in any definite way—however, the course of the ship may be corrected through the observation of the stars by a conscious being. The subject can even link things in time, not only in space: events separated by billions of years become related in consciousness. But the most important consequence of the universality of subjective mediation is the subject's ability to link not only material things but also any aspects of their existence, abstracted from the things themselves. Relations between things are linked to things, or other relations, and there is no limit to the complexity of such abstract mediations. The subject is the only way to establish the links like that, and it is for that universal mediation that consciousness appears in the world.

As universal mediation, a subject can also mediate any relations between subjects, including the subject's relatedness to itself. This representedness of the subject's activity in the subject is known as *consciousness*.

It should be stressed that a subject cannot be reduced to a thing, or an organism, though both inanimate things and living things are necessary to represent any particular subject. That is, consciousness cannot merely be a property of the human body, or any part of it.

Like on the other levels of mediation, there is a hierarchy of subjectivity: any group of subjects can form a higher level subject, which allows for much more diversity than in biological communities.

While each thing can be linked by the subject to any other thing, the subject primarily links it to itself: $X \to S \to X$. In other words, once the thing has been assimilated by the subject, it will contain the subject inside, thus becoming an *object*. That is, an object can be defined as anything at all in its relation to the subject. For the subject, the world unfolds itself as a hierarchy of objects, *nature*.

This means that the subject mostly mediates the relations between objects, rather than mere things or their properties, and the mediation scheme takes could be rewritten as $O \to S \to O'$. On the highest level, the subject presents itself as the universal way of nature's self-reproduction; this universal reflectivity of nature is called *spirit*.

It should be stressed, that the subject is an object too, and hence a part of nature. However, this is a very special kind of object, namely, performing the universal mediation. The things and organisms implementing the subject do not need to be representatives of the subject in every respect, and they retain their existence as material things, or living creatures. The subject should not be identified with its implementation, and no finite implementation can represent subjectivity in full. For instance, any human combines being a material thing with being a representative of a biological species, and, in certain respects, the subject of conscious activity.

Finally, we have the triad of mediator types:

$$a\ thing \to an\ organism \to a\ subject\ ,$$

which corresponds to the hierarchy of universality of mediation:

$$randomness \to necessity \to freedom\ .$$

As usual, any hierarchical structure is relative, and the living creatures could not have existed if there were nothing in the inanimate world that would be akin to life; similarly subjectivity has its origin in life, and reflected in it. Within the random mediation (characteristic of the level of existence), there may be a hierarchy of specific forms, differing in their universality. For instance, sequencing in atomic and chemical reactions could be considered as a germ of life. On the level of consciousness, there is a hierarchy of its manifestations, some of which may resemble animals, or even inanimate reflection. It is only in a specific context that the attribution of a certain type of behavior to consciousness will have sense.

## *The subject and activity*

The world's universal relation to itself (reflection) includes the world's self-reproduction. The inanimate, animate and conscious levels of reflection are characterized by their own place in that reproduction. Conscious production of things and their relations is called *activity*.

On the inanimate level, mediation is basically identical to interaction, and one body is merely absorbed by another, possibly changing its state. No further interactions are assumed. The behavior of a living creature only passively influences its environment: an organism may cause an environmental change as a *side effect* of its behavior, which is not intended to be used by that organism in the future. The activity of the subject implies transformation of the objects involved, their adjustment to the subject's ways. In any activity, some changes in the world are *produced* to be used by the subject, though not necessarily the producer. Such artificial objects, produced by the subject for the subject, are called *products*, and a product can be considered as a synthesis of an object and a subject possessing the attributes of the both. No product can exist otherwise than in its relation to the subject; taken apart from a definite culture, no thing can be considered as a product of conscious activity, remaining a mere thing.

Every object is a product, since its definition implies relatedness to the subject. This is a consequence of the presence of both the material and ideal sides in any reality. In particular, the subject as a special kind of object is a product too. Both objects and subjects are repeatedly reproduced in the two ways, or aspects. The primary object cycle can be expressed with the scheme:

$$... \to O \to S \to O' \to ...$$

This representation puts stress on the reproduction of the world as the result of conscious activity. It is one of the determinative traits of the subject, to produce objects rather than merely perceive them. The complementary subject cycle is the inverted form of the object cycle:

$$... \to S \to O \to S' \to ...$$

Here, the stress is on the subject's development through interaction with the world. Any activity is the unity of the both objective and subjective reproduction. As a special case of mediation, each particular activity can be represented by its product.

In the continuous train of activity, one could distinguish individual acts of reproduction, which may be of either $O \to S \to O'$ or $S \to O \to S'$ type. The former triad represents individual *actions*, while the latter scheme $S \to O \to S'$ describes *communication acts*, or *transactions*. Any activity includes both action and communication. However, the same abstract scheme can represent quite different kinds of action or communication. Thus, the triad $S \to O \to S'$ can refer to either communication of two individuals, or communication of a person with him- or herself, interrelations between a person and a group, processes of personality development *etc*. Similarly, the action scheme $O \to S \to O'$ can span the range from a routine operation to the development of the world mediated by conscious activity. Virtually, due to the universality of subjective mediation, the entire world is to become involved in conscious activity and adapted to the subject's needs, cultivated. In this way, the very distinction of the object, the subject and the product will be lifted up, so that the whole world in its entirety will become the object, the subject and the product.

*Reproduction and creativity*

The material side of activity is in rearrangement of the world, assimilating and re-creating it. People use objects to make products, and any product is intended to become an object for some other people, in their own activity. The general scheme of such subject-mediated reproduction, $O \to S \to P$, combines two complementary acts, *consumption* and *production*, expressed by the links $O \to S$ and $S \to P$ respectively.[67]

Objects and products can be very different, being in no way restricted to material things; in particular, reproduction of the world by the subject includes reproduction of reflection, on any level. For instance an individual act $O \to S \to P$ can correspond to using $O$ as raw material for producing $P$; the same scheme with $O$ as a social relation describes one's behavior as dictated by moral norms or cultural stereotypes. The product $P$ can be either a tangible thing or a sublime change in one's soul, or in people's relations. In any activity, the subject reproduces both nature and subjectivity itself.

On the syncretic level, consumption and production are the aspects of the same act. Thus, writing a letter on a sheet of paper, we spend some ink; satisfying hunger, we consume food; attending a ballet show (consumption), we produce certain mental structures inside us. In other respects, production and consumption can be formally separated, with many acts of consumption accumulated for a single act of production, and a single act of consumption leading to different products. This is the analytical level, where consumption physically precedes production. Syncretism is lifted up on this level of activity, and any one of the actions comprising it can be considered as a simple unity of consumption and production, with the favorable conditions for further actions as partial products. On the synthetic level, all the analytical activities are included in the integral process of cultural reproduction, restoring the objective and subjective conditions for each specific activity.

In the object reproduction cycle $O \to S \to O$, the same objects become produced again and again, to allow repeated activities. This special form of the cycle expresses *simple reproduction*, and it is the way human culture is conserved despite all its inherent dynamics. However, there is a complementary interpretation of this scheme, stressing that every act of production will change the world, introduce something that did not exist before, so that the new product would be qualitatively different from any previously known object. This aspect of reproduction is known as *creativity*. In general, since the product is a synthesis of the object and the subject, it is different from mere object by definition. However, in every particular act of

---

[67] From the dialectical viewpoint, $O$ is the thesis, $P$ is the antitheses, and $P$ is the synthesis of the two opposites $O$ and $S$.

production, the creative component may be of different importance. Still, creative work is the attribute of the subject, and one of the criteria of subjectivity and consciousness.

In the picture of activity as cyclic reproduction of both the objective and the subjective sides of the subject/object interaction, one could formally unfold (mediate) the arrows in the scheme $O \to S \to O'$, which gives

$$O \to P \to S \to P' \to O'.$$

Thus, the influence of the object $O$ onto the subject $S$ (or the object's assimilation by the subject) becomes mediated by a special product $P$ serving as an *instrument*; similarly, the subject $S$ influences (produces) object $O'$ using a *tool* $P'$. Occupying the place of a mediator between the subject and the object, every tool can be characterized from both objective and subjective sides, that is, its functioning in the physical world and the modes of the tool's usage; also, an instrument combines transmission of the world's influence on the subject with subjective filtering, selecting the relations relevant to the current activity. This could be represented by the scheme

$$O \to (o \leftrightarrow s) \to S \to (s' \leftrightarrow o') \to O',$$

or, in a converted form,

$$(O \to o) \to (s \to S \to s') \to (o' \to O').$$

In other words, when involved in conscious activity, objects present themselves to the subject in a very special manner; complementarily, tool/instrument mediated activity makes the subject expand, assimilating a part of the external world and developing a kind of an *inorganic body*. In its ultimate development the subject will embrace the entire world, becoming identical to it.

*Communication*

In communication, $S \to O \to S'$, two subjects cooperate within a common activity. This scheme does not make any sense on itself, as if the subject created the world from nothing and then absorbed it back.[68] The origin of communication is from repeated activity, in which any subject can be substituted by any other:

$$O \to S \to O' \to S' \to P$$

---

[68] It was the basic scheme of the development of the objective spirit in Hegel's philosophy. Subjective idealism entirely eliminates the object, reducing this scheme to subject's self-communication: $S \Rightarrow S'$.

This scheme describes the substitution of ione subject by another within a definite activity, their interchangeability in this respect. Instead of directly producing the product $P$, the first subject $S$ only produces the condition for another subject $S'$ to produce the final result using the intermediate product $O'$. Thus people manipulate other people, using them as their tools to achieve what is not directly achievable. The possibility and necessity of such a second-order (subject mediated) production is an immediate consequence of the universality of subjective mediation. Higher animals can develop primitive forms of manipulation, but it is only with conscious beings that the majority of behavioral acts involve manipulation. However, if such a direct manipulation dominates in one's activity, this is a sign of primitive consciousness; self-consciousness and reason primarily demand well-developed *self-manipulation* through the other people and the society as a whole. This becomes possible because of reflexive communication, as described by the scheme

$$O \to (S \to O' \to S' \to O'' \to S) \to P$$

For an external observer, this looks like ordinary production, $O \to S \to P$, with the only difference that there is a delay between consuming the object $O$ and producing the product $P$, and in the middle the subject communicates with other subjects in order to get prepared for the final production. As one can see, this is equivalent to self-communication through somebody else's activity:

$$O \to (S \to (O' \to S' \to O'') \to S) \to P,$$

or

$$O \to (S \to O(S') \to S) \to P.$$

This is how people come to communicating with themselves, self-communication. Every instance of such communication is interiorized communication with the others, with the society in general.

It is important that different subjects perceive each other as activities, and not mere objects. Thus, in the scheme $S \to S' \to P$, the subject $S$ is yet another object, with no subjective quality; on the contrary, in the scheme

$$(O \to S \to O') \to S' \to P,$$

which is yet another position of the scheme used above to illustrate the origin of communication, it is the whole activity $O \to S \to O'$ that plays the role of the object for $S'$, and hence $S$ is now perceived as a subject, as universal mediator.

The demand of universality implies that there must be an object mediating communication in a universal way. This universal mediator of communication is known as *language* (hereafter denoted by the letter *L*). It does not need to be the common verbal conversation; the idea of language is wider, including both verbal and non-verbal components, as long as they are used in a universal manner. Thus, words can be mere voice signals, without the specifically human cultural reference; similarly, silent communion can convey much more conscious content than vast prolixity. Almost any object can be used in a language-like mode, though some objects are more convenient for universal mediation than the others. Speech became the basis of human communication due to its high versatility in the conditions of the planet Earth; this does not mean that it is the only possible implementation of language, and that, some day, it would not lose its dominance even in humans. The overall direction of development is towards more diversity of communication, increasing its non-verbal component.

Like in objective reproduction, communication can be split into the separate acts of *expression*, $S \to L$, and *attending*, $L \to S'$. In indirect communication, expression and attending are separated in time, with some objective process, or activity, proceeding between them:

$$S \to (L \to L') \to S',$$

or

$$S \to (L \to S'' \to L') \to S'.$$

Quite often, the role of the intermediate subject in communication is played by a higher-level subject (a social layer, a class, the society as a whole). In particular, words are always perceived in the current cultural context, saturated with some "common sense" as one of the relevant connotations. This is why the same words can mean opposite things, when said by different people referring to different cultural context.

*Inner activity*

In the cycle $S \to O \to S'$, one can formally fold objective mediation and consider "purely subjective" development:

$$\ldots \to S \to S' \to S'' \to \ldots$$

In such a scheme, the subject seems to develop entirely through communication and self-communication. In the latter case, the scheme $S \to S' \to S$ would express an activity-like process *within* the subject, substituting the subject (or rather one of its objectivated forms) for objective mediation. This apparently objectless process is known as *inner activity*, and it is a complement of the *outer*, object mediated activity.

In outer activity, the subject produces some object that is to be consumed by this or another subject to maintain the active existence. In inner activity, the subject itself plays the role of this intermediate product; the subject becomes an object for itself. This reflexivity results in growing of a hierarchy within the subject; its outer consequence is a hierarchy of subjects.

The same scheme of inner activity applies to the development of subjectivity in general, as a level of reflection; also, it describes the formation of the various forms of the collective subject (groups, societies). The scheme of inner activity will equally refer to psychological processes in the personality, to the relations between the members of a group, to the interaction of social forces. As an inter-level process, it describes the development of individuals through society (socialization). Individual subjects do not need to be physical individuals, humans; they can as well be groups of people joined by some common activity into a whole, or even imaginary characters.[69] As soon as such a collective formation can mediate (at least indirectly, through other people) relations between things and other subjects, it can be called a subject, and it can develop regardless of the presence of the physical body.

The origin of inner activity is in the intrinsic subjectivity of any object, which is always a product and hence reflects some aspects of the subject. Therefore, any objective mediation of the inter-subject relations $S \to O \to S''$ is always *activity mediated*:

$$S \to \begin{Bmatrix} S' \\ \updownarrow \\ O' \end{Bmatrix} \to S''$$

which includes the component $S \to S' \to S''$. Due to the universality of subjective mediation, the mediating subject $S'$ can become a part of $S$; if the same subject $S$ is substituted for $S''$, the outer sequence $S \to S' \to S''$ becomes reproduced within the same subject, which is known as *interiorization*. Depending on the level of the subject (an individual, a collective subject, the society), interiorization takes different forms. For instance, one could distinguish phylogenic and ontogenetic interiorization,

---

[69] The example of Kozma Prutkov could be mentioned in this context. Three Russian poets worked under that alias, creating in a style different from their own, so that no one of them could be associated with the bright personality they produced.

through historical development and learning, through cultural influence and education.

Inner activity can also be understood as expansion of the subject's non-organic body. Interacting with the world through instruments and tools, the subject includes their subject related aspects (modes of operation, typical applications *etc*) into its internal hierarchy, which leads to a scheme like $s \to S \to s'$, with $s$ and $s'$ denoting the parts of the instrument and tool included in the subject. In this scheme, both the object and the product are within the subject, which explains the term "inner activity".

Since no inter-subject relations can exist outside the cycle of reproduction of the objective world, every inner activity originates from some outer activity, and it can be unfolded in outer activity under special circumstances. This process of *exteriorization* is an important element of cultural inheritance, since it provides a mechanism for new subject formation, when new subjects first appear as an inter-subject relation. For instance, when the parents give birth to a child, they form a specific relation between them long before the physical birth, and, sometimes, even before the conception. This inner relation becomes exteriorized and projected onto the newly born organic body, through including it in the process of objective reproduction and social relations.

## Inside the Subject

### *Unfolding mediation*

When an object $M$ mediates the link between some other objects, $X \to M \to X'$, it turns its different sides to the objects $X$ and $X'$. For $X$, it manifests itself as a specific object $M(X)$, that accepts the influence of $X$ and is transformed under its influence; for $X'$, the same object $M$ appears to be a quite different object $M(X')$, acting upon $X'$ and transforming it. This means that the mediator $M$ must be able to behave in two different (and even opposite) ways. Typically, this implies the existence of some inner structures $S$ and $R$, that represent the reflection of $X$ in $M$ and the possibility of $M$ influencing $X'$, respectively. Following the general rule of dialectical logic, we conclude that the necessity of simultaneously being two different objects $M(X)$ and $M(X')$ constitutes the dialectical contradiction in $M$ leading to its transformation and development. This development is to follow a number of universal stages: syncretism, analytical separation of different aspects, and then their synthesis at a higher level.

In the simplest case, the two aspects of the mediator are often merged with each other and inseparable. The same thing in the same time accepts influence and transmits it to another thing. Such *syncretic reactivity* could be expressed by the scheme

$$X \to (SR) \to X'.$$

The structure S is built of the same elements as the structure $R$; however, in general, $S$ does not need to coincide with $R$; that is, the mediator $M$ can not only mediate the relation between $X$ and $X'$, but also influence $X'$ on its own, as an independent object, regardless of any mediation.[70]

Further development of this difference leads to the separation of the two sides of the mediator, so that accepting the influence from outside and influencing other objects are implemented in separate structures in $M$:

$$X \to (S \to R) \to X'.$$

Now, the projection $S$ of $X$ into $M$ is implemented differently from the prototype $R$ of $X'$ in $M$, and there must be some inner process to transform $S$ into $R$. On this level, the designations $S$ and $R$ already refer to the (states of the) respective subsystems in $M$ rather than to the different aspects of its existence. However, at this level, the transformation of $S$ into $R$ is still immediate, either rigidly pre-defined ("hard-wired") or random.

At a higher level, the connection of $S$ to $R$ becomes internally mediated by an internal structure $C$, so that the inner motion in $M$ would reproduce the entire act of mediation:

$$X \to (S \to C \to R) \to X'.$$

Thus the mediator $M$ becomes a complete system, with input $S$, output $R$ and inner state $C$. This is how any object can be unfolded in a hierarchical structure. Obviously, repeating this logical scheme, one can get deeper in the core of the mediator, discovering a hierarchy of inner mediations:

---

[70] One could mathematically model this level of mediation with the composition of two set mappings $\mu: X \to M$ and $\mu': M \to X'$, so that some element $x \in X$ is first transformed into element $m$ of the set $M$, and then $m$ is transformed into element $x' \in X'$: $x \to m \to x'$. The element $m$ is in the same time the image of $x$ in the mapping $x \to m$ and the original of $x'$ in the mapping $m \to x'$. The image of the whole $X$ in $M$ under such a transformation is denoted as $S = \mu(X)$, while the prototype of $X'$ in $M$ is denoted as $R = (\mu')^{-1}(X')$. If $S$ coincides with $R$ then there is a direct mapping of $X$ into $X'$, and the composition becomes trivial.

$$X \to (S_0 \to (S_1 \to C_1 \to R_1) \to R_0) \to X',$$
$$X \to (S_0 \to (S_1 \to (S_2 \to C_2 \to R_2) \to R_1) \to R_0) \to X' \; etc$$

In every particular study, it is important to adjust the level of consideration to the practical needs, keeping in the mind that a different approach may be required in a different situation.

In the above schemes, all the inner levels can refer to any level of reflection; similar hierarchies can develop in inanimate matter, in biological systems, or in conscious beings. The meaning of the schemes will be different, depending on the type of mediation. The scheme of inner mediation $S \to C \to R$ only says that reflection of the external world transforms into an outer action through a material process that is localized *within* $M$ if the level of distinction between the inner and the outer is properly chosen.

The logic of inner unfolding complements the already described logic of expansion through assimilating a part of environment:

$$O \to M \to O',$$
$$O \to (o \leftrightarrow m) \to M \to (m' \leftrightarrow o') \to O',$$
$$(O \to o) \to (m \to M \to m') \to (o' \to O').$$

The distinction between the two logics is relative, depending on the context. On the lower levels of the hierarchy of reflection, inner unfolding prevails, while the conscious subject develops almost all of its inner hierarchies through outer expansion.

*Inanimate world*

The unity of the world implies that all the levels of mediation must be present on the physical level too, for consciousness to be able to originate from it. However, the randomness of mediation characteristic of this level will assimilate the higher levels of mediation to the lower, so that syncretic mediation would dominate in this position of the hierarchy.

There are numerous examples of syncretic mediation in physics, chemistry and any other sciences studying the physical world. For instance, in mechanical devices, one can consider their parts as rigid, transmitting any motion from one end to another in no time. When you turn a key in a lock, and your effort is transmitted to the lock mechanism, to open or close the lock. Or, in a mechanical watch, one gear turns another through a number of intermediate gears. In all these cases, the changes in the interacting bodies themselves are negligible, and it is only their motion that will visibly change. Similarly, in the thermodynamics of the ideal gas, or in hydraulics, the medium serves to convey energy and transform it from one form to another. There is no distinction between the parts of the medium, it works as a whole. In chemical reactions, water, or other medium, can simply carry one reacting substance to another, without being involved in the reaction itself. In catalysis, the intermediate agent first binds the incident substances, and then releases them in a different combination. In inanimate nature, this kind of mediation does not much differ from mechanical transport of the reagents, the catalyst only serving to put the molecule in the suitable position for reaction.

Systems of any complexity can be built from such syncretic elements. However, in nature, all such combinations are essentially random and therefore rather limited in both their scope and complexity. It is only through human activity that the majority of the possible inanimate systems can be produced.

In physical systems built from syncretic elements, the elements themselves do not significantly change due to their involvement in the system. According to the general principle of the hierarchical approach, such external complexity can also, on a different level of hierarchy, manifest itself as inner complexity. Thus, in quantum electrodynamics, one finds that a photon, while transporting the electromagnetic interaction from one electron to another, interacts with electrodynamic vacuum, producing virtual electron-positron pairs that instantly disappear, still influencing physical interactions. In atomic physics, an atom can be directly ionized by ultra short wave radiation; the same result can also be produced trough first absorption of a photon with the atom excited to an autoionizing state, and then emitting an electron, thus discharging excitation. This sequence is different from mere cascade reaction, which is a chain of reactions that do not depend on each other; in a *virtual* reaction, all the intermediate products do not exist for the observer, and one reaction channel cannot be separated from another (which is known as quantum interference). Rather complex inner hierarchies can arise in that way. Nevertheless, for the external world, only the final outcome matters and the overall syncretism of the physical level is never violated.

*Organic life*

Unlike random physical interactions, life is the realm of necessity, implying quite definite sequences of mediations for each organic form to keep its live existence. Organic life is based on chemical cycles, chains of reactions that repeatedly produce nearly the same combination of substances in state in a similar state. An organism consumes the necessary building blocks and energy from the environment, and returns the wastes into it. This resembles catalysis in chemistry, with the difference that the organism itself keeps changing in that metabolic process—it grows, matures, ages and dies.

At any biological level, an organism is not detachable from its environment, passively depending on the supply of the necessary materials. The organism reproduces itself, provided there are favorable conditions, but it does not reproduce the conditions themselves. It is only in a symbiosis of many organisms (ecosystem) that the reproduction of different organisms can form a relatively stable cycle, supporting all its members.

In the most primitive forms, one finds organic mediation as the syncretic unity of simple irritability $S$ and spontaneous activity $R$: $X \to (SR) \to X'$. However, most organisms develop special organs (sensors) for accepting irritation and transforming it into signals that can activate special effectors. On a definite level of development, such a commutation is performed by the nervous system, a group of cells that are specialized in accepting, transforming and distributing activation between the different subsystems of the organism. In the scheme $X \to (S \to R) \to X'$, the reflection of the environment in the organism $S$ is usually called a *stimulus*, while the structure of the activation of the organism's effectors $R$ is known as *reaction*; the standard technique of associating stimuli with reactions in living organisms is *reflex*.

In a well developed form, a reflex can commute stimuli to reaction in an internally mediated way, as described by the scheme

$$X \to (S \to C \to R) \to X'.$$

In an inborn reflex, the mediator structure $C$ is formed together with the organism; in a conditioned reflex, it can be formed dynamically, reconfiguring neural activation patterns rather than changing the cerebral structures. This, however, does not modify the basic mechanism of biological mediation, since the behavioral patterns available to an animal are determined by its biological body, and no essentially new reactions can be expected. Like on the inanimate level, the animal's being involved in conscious activity can develop organic and behavioral forms that can never develop in nature. Such mediation sequences can only be based on the available physiological mechanisms, and, once formed, they become as rigid as any other reflex.

In the development of inner complexity in animals, there are two complementary processes. Primarily, generalization expands the range of stimuli associated with a specific reaction. Schematically, there are two slightly different mechanisms behind this capacity. Thus, since the number of the possible internal states is limited in lower animals, different stimuli $S_1$ and $S_2$ cannot be properly differentiated, leading to the same internal state $C$, and hence to the same reaction $R$. In higher animals, the complexity of inner states allows better representation of the world, but lack of differentiated enough reactions means that different internal states $C_1$ and $C_2$ can lead to the same reaction $R$, so that the sequences of inner mediation $S_1 \to C_1 \to R$ and $S_2 \to C_2 \to R$ are equally possible. These two mechanisms are not isolated from each other. As a rule, conditioning will first produced a much generalized reaction, which is due to the syncretic nature of primary reflection in animals, that is, the most primitive biological aspects of the situations are first discerned, and the most basic internal processes become initiated. The limited range of reactions limits the fineness of primary signal discrimination.

In the reverse process of specification, stimuli become differentiated on the basis of a common property, to produce different reactions. Higher animals can differentiate almost identical stimuli, but they can only express their recognition using the available means, which are not always adequate. In nature, mainly the gross assessment of the situation is of adaptive importance, and too fine details do not matter; the ability to discriminate them originally appears as a by-product of biological evolution. However, high sensitivity can become adaptive in complex ecosystems, where the same reaction can have different consequences under different circumstances. Thus the variability of the environment compensates for organic deficiencies. If the animal lives in contact with humans, its environment becomes extremely diverse, and the animal can realize its ability of reflex specification in full. However, this can only happen when humans pay attention to the different modes of animal behavior and their context.

*Subjectivity*

The universality of subjective mediation results in that the whole world is reflected in the organization of the subject, including the subject and conscious activity. This universal reflection is due to the world's transformation by conscious beings and regarding any thing as a product, rather than a thing on itself. While an animal depends on the environment, conscious beings re-create their environment to eliminate rigid necessity and achieve freedom. This means that self-reconstruction is an attribute of any kind of the subject; in particular, all the inner structures are no longer permanent, and anything in the subject can be intentionally changed, if necessary. But, since syncretism, analysis and synthesis are the universal stages of development, subjectivity can manifest itself on either of the inner mediation levels: $(SR)$, $S \to R$ or $S \to C \to R$ in any inner or outer activity, folding and unfolding it to an adequate degree.

An object is defined in its relation to the subject; conversely, it is only objects that the subject can perceive. Consequently, in the subject, every stimulus will contain a subjective component, which implies self-reflection. In a person, this self-reflection is primarily reflection of the others as subjects; its origin is in the reflection of the subject's own activity. That is, in subjective mediation, a stimulus $S$ is a hierarchy of the person's relations to the others, and the society in general, taken in a specific respect, in relation to an object. Similarly, the inner state $C$ and reaction $R$ become hierarchical, being correlated with all the levels of the subject.

In the chain of actions performed by the same subject

$$O \to S \to P_{12} \to S \to P_{23} \to S \to P,$$

with intermediate products $P_{12}$ and $P_{23}$, one can represent every external act as mediated by an inner process in the subject:

$$O \to (S_1 \to C_1 \to R_1) \to P_{12} \to (S_2 \to C_2 \to R_2) \to P_{23} \to (S_3 \to C_3 \to R_3) \to P$$

To produce inner hierarchies, this scheme can be folded differently. Thus, if $S_{1,2,3}$, $C_{1,2,3}$ and $R_{1,2,3}$ are of the same kind, they can be considered as the levels of hierarchy in stimulus $S$, internal state of the subject $C$ and reaction $R$, respectively:

$$O \to \left\{ \begin{pmatrix} S_3 \\ S_2 \\ S_1 \end{pmatrix} \to \begin{pmatrix} C_3 \\ C_2 \\ C_1 \end{pmatrix} \to \begin{pmatrix} R_3 \\ R_2 \\ R_1 \end{pmatrix} \right\} \to P$$

In this process, the object becomes hierarchical as well, $O = \{O_1, O_2, O_3\}$, and hierarchical relations to the subject simultaneously on multiple levels.[71]

The subject's ability to perceive the world and act simultaneously on different levels is one of important consequences of the universal reflexivity characteristic of consciousness. In particular, this many-aspect and multilevel

---

[71] Such a scenario also implies that the products $P_{12,23}$ are of the same kind as the object $O$ (being its aspects or features), which describes the process of extended reproduction of the object in human activity. Formally, lift-up $P_{12,23} \to O$ is implied after each action, that is, extraction of the objective side of activity, its embodiment in the culture.

organization distinguishes the inner structures of conscious beings from the "flat" perception and behavior of the animals. Even though a chimpanzee can form inner mediation chains of up to thirteen (and probably more) phases, it cannot fold them in an inner hierarchy, making them a single act. Similarly, the fantastic memory of a chukchi reindeer breeder does not make him any more conscious than an absent-minded European scientist. Intelligence does not imply reason.

In an alternative representation, the object $O$ and product $P$ remain the same but their relation is indirect, mediated by a hierarchical inner activity:

$$(S_3 \to C_3 \to R_3) \to P$$
$$\uparrow$$
$$S_2 \to C_2 \to R_2$$
$$\uparrow$$
$$O \to S_1 \to C_1 \to R_1$$

In this case, $S_2$ will be qualitatively different from $S_1$, reflecting the very process $O \to (S_1 \to C_1 \to R_1) \to P_{12}$, rather than its result, which requires a different kind of lift-up: $P_{12,23} \to S$, accentuating the subjective side of activity. Such a scheme is useful to represent the extended reproduction of the subject in conscious activity.

Phylogenically, the development of subjectivity as such is thus described. The root of this development is in the growing complexity of the inner activity mediating every act of outer mediation. For yet another aspect, the social relations become ever more complicated, hierarchical and indirect.

Applied to individual development, the same scheme describes learning and education, socialization and assimilation the achievements of a given culture. Thus the history of phylogenic development is reflected in one's individual history.

The two schemes of activity folding in the inner hierarchy of the subject describe the two components of any subjective development, structural growth due to development of the cultural environment and developing a more complex functionality through participation in existing activities. Complementing each other, they are both based on the reconstruction of the world by the subject, and the reflection of the changes thus made in the organization of the subject itself. Unlike in the animal world, such a reflection is possible within the life cycle of an individual subject, and not only through biological selection in phylogeny. While biological laws remain applicable to the organic body of the individual carriers of subjectivity, biological evolution is no longer determining the directions of organic

# ONTOLOGY

development, which is much more influenced by the cultural factors. For the individual subject, this looks as if there were no external world independent of the subject's existence, and all the inner development were due to mere exercising the free will of the subject. Such an impression is a correct reflection of the fact that subjectivity does not exist inside any organism, it is always in inter-individual relations, which, under certain social conditions, can be alienated from the individual and countervail him as a separate entity.

## *Mental processes*

Considering the processes inside an individual subject, one must always relate them to the outer activity, and communication with the other subjects, both on the same and other levels. Any inner activity originates from outer activity and communication, and the inner organization of the subject, and every inner process, reflects the organization of the culture. The very existence of the subject is due to the repeated reproduction of the cultural phenomena in a hierarchy of activities. However, subjectively, this cycle manifests itself through repeated reproduction of the subject's inner activities as typical sensory and motor structures[72] and the typical chains of mediation between them:

$$... \to S \to C \to R \to S' \to C' \to R' \to S'' \to C'' \to R'' \to ...$$

When $S'$ (or $S''$) is of the same kind as $S$, this chain forms an inner cycle:

which resembles the usual feedback schemes in systems theory, with the output of the system tied to its input. Here, the key link $R \to S$ is intrinsically culture-dependent, as it lifts up objective mediation (and productive activity, the reorganization of the world to satisfy the subject's needs), which results in apparently direct transformation of the reaction $R$ into sensory input $S$—an inverse of the usual stimulus-reaction sequence. It is only in the society of conscious beings that such a folding can become regular, which is yet

---

[72] When applied to the subject, the attribution of the structures $S$ and $R$ to the sensor and motor aspects of activity is rather figurative than literal, since the subject is not contained in an organic body, and hence any inner structure or process is rather a cultural phenomenon than a physiological act.

another aspect of the subject's universality. Since the sensory structure $S$ is normally an image of the world, and the motor structure $R$ is directly related to the subject's action, the projection of $R$ onto $S$ must be related to the ability of *imagination*.[73] Indeed, any human fantasy is always a prediction of the change in the world that would be caused by one's action; even though people's dreams may seem predominantly passive, they still imply the dreamer's participation, at least on the level of mere observation.

One cannot immediately observe the inner structures $S$, $C$ and $R$; however, since the stimuli and reaction are culturally standardized, the structures $S$ and $R$ can be derived from the current activity and the cultural stereotypes associated with it.[74] On the contrary, explication of the inner mediation $C$ requires a rather complex investigation, and in most cases this structure is only assumed; for the outer observer, there is an apparently direct connection $S \Rightarrow R$, which looks like a simple reflex, but is different in that it remains culturally mediated and hence universal. Such virtual mapping of $S$ onto $R$ abstracted from the inner mediation $C$ can be identified with a *mental act*.

Combining the links $R \to S$ and $S \Rightarrow R$, one arrives to $S$ and $R$ reproducing themselves through each other:

$$\ldots \to S \Rightarrow R \to S \Rightarrow R \to \ldots$$

This inner activity is called a *mental process*. Similarly to reproduction of the subject and the object in an outer activity, through mutual reflection and mutual penetration, the interaction of the distinct sides of the subject leads to their development through each other.

Also, employing the cyclic nature of any inner activity, one can lift up mediation in the sequences $R \to S \to C$ and $C \to R \to S$, arriving to mental acts $R \Rightarrow C$ and $C \Rightarrow S$. The corresponding mental processes, obtained trough combining these secondary links with the primary links $C \to R$ and $S \to C$ respectively, are represented by the schemes

$$\ldots \to R \Rightarrow C \to R \Rightarrow C \to \ldots$$

and

$$\ldots \to C \Rightarrow S \to C \Rightarrow S \to \ldots$$

To summarize, there are three kinds of mental acts ($S \Rightarrow R$, $R \Rightarrow C$, and

---

[73] T. Ribot *Essai sur l'imagination créatrice* (Paris, 1900)

[74] This is like quantum physics derives microscopic structures from macroscopic observations imposing asymptotic constraints determined by the organization of the experiment.

$C \Rightarrow S$), forming the secondary cycle of inner activity:

The relations between the inner structures of the subject are reversed in the secondary cycle. This is how they are often presented to the subject itself. Thus, the subject's reactions $R$ seem to influence the subject's internal state $C$, while, in the primary sequence, the internal state $C$ causes certain reactions $R$; also, the subject is apt to suppose that at least a part of stimulation $S$ comes from inside, $C \Rightarrow S$, while the primary dependence is reverse: external stimulation influences the inner states of the subject, $S \to C$.

The schemes of inner activity allow for numerous interpretations. Thus, the scheme $S \Rightarrow R \to S$ can refer to the development of the subjective image of the world $S$ via explorative behavior $R$, which is an obvious correlate of a *cognitive process* in the classical psychology. On the conscious level, the link $C \Rightarrow S$ can be identified with an act of planning, while the complementary link $R \Rightarrow C$ corresponds to self-control. The corresponding schemes of inner activity, $R \Rightarrow C \to R$ and $C \Rightarrow S \to C$, describe, respectively, consciously constructing one's actions (the *will*) and the evolution of one's inner state through re-interpretation of an external stimulus (the ability of *feeling*), which gives the other two members of the well-known psychological triad: cognition, affects, volition.

The existence of the three types of mental processes is evidently related to the distinction of the structures $S$, $C$ and $R$ in any subjective mediation. Each class of mental processes can be considered as an alternative representation of a corresponding structure: $S$ for cognition, $C$ for feelings, $R$ for volition. This attribution, which is reflected in the common characterization of mental processes, helps to accentuate the specific quality of each class, indicating there is a primary form of each metal act, like the primary sequence $O \to S \to O'$ and the secondary sequence $S \to O \to S'$ in the infinite cycle of the subject-object reproduction (activity):

$$\ldots \to O \to S \to O' \to S' \to O'' \to \ldots$$

In the primary form of a scheme of a mental process, the primary link ($\to$) is on the higher level, and hence the mental act (the secondary link, $\Rightarrow$) is subordinate, serving to produce something "material". In the secondary form, conversely, the accent is on the reproduction of the mental act itself.

The very existence of mental processes as the subject's inner motion is due to hierarchical inner mediation, the subject's self-communication, inner activity. All the mental processes are only different manifestations of the same inner activity, and consequently, there can be no "pure" cognition, feeling or will. In every particular activity, one class of mental processes comes to the top only to give way to another, in another position of the hierarchy. The dynamics of this conversion accounts for all the complexity of a conscious act.

*Cognition*

The cognitive mental process is described by the scheme $... \to S \Rightarrow R \to S \Rightarrow R \to ...$, where the image of the world initiates certain actions, which modify the original image and cause the reproduction of the same activity on a higher level. Since both structures involved in a cognitive process, $S$ and $R$, can be derived from the standard templates of activity assumed by the particular culture, cognition is easier to study than the other types of mental processes. That is why, for centuries, most psychophysical research was concentrated on the cognitive operations, and there was even a tendency to reduce all the subjectivity to cognition, like in the philosophy of European rationalism (Descartes, Leibniz).

In cognitive processes, one's sensations ($S$) appear to be immediately produced by one's own intentions ($R$); this illusion lies in the foundation of numerous idealistic theories of consciousness. From the earliest times, people imagined powerful sorcerers who can do all kinds of miracles by mere effort of will; in a vain attempt to transform such fantasies into reality, people practiced various magic rites. In modern idealistic philosophy, this standpoint has been refined to complete solipsism, when the entire world is thought to be one's imagination, a dream. Despite of the obvious absurdity of such a position, many people still advocate some of its weaker (that is, inconsistent and eclectic) varieties. For a bourgeois philosopher, idealistic ideas are very attractive, since they perfectly correspond to their experience of a privileged person, whose will is normally fulfilled by the others.

There are two principal schools in the idealistic philosophy of cognition. One of them puts stress on the subject's intentionality and identifies the subject with the $R$ structure, while the opposite tendency is to reduce the subject to its passive aspect, to the image of the world. Both trends implicitly attempt to project the normal scheme of subject-object relations $O \to S \to P$ (the object, the subject, the product) onto the scheme of the cognitive process, which can take two opposite positions, $S \Rightarrow R \to S$ and $R \to S \Rightarrow R$.

The former possibility presents the world as mere imprint of the subject's will; the latter, conversely, pictures it as an active substance, determining the subject's "flow of consciousness".

However, the apparent independence of cognition from the objective world is only illusionary. Both components of the cognitive process, $S \Rightarrow R$ and $R \to S$, are mediated in an essentially material way, by the objective organization of the subject and the culture, respectively. And both kinds of mediation are necessary for a cognitive process to unfold. Consequently, cognition cannot be understood exclusively on the basis of the knowledge of cerebral functionality (as in the so called cognitive science), nor can it be explained solely by cultural influences (cultural psychology).

In the scheme of the cognitive process,

$$\to S \Rightarrow R \to S \Rightarrow R \to \ldots,$$

one can consider the subjective structures $S$ and $R$ in different aspects, thus obtaining different interpretations of the same scheme. Typically, all the components of the scheme are understood as the states of the same individual; however, this individual cognition is not the only possibility, and one can, for instance, describe cognition as a process of communicating the image of the world and intentions from one individual to another, and hence creating a collective picture of the world, as well as collective interests. If the structures $S$ and $R$ belong to the different levels of the subject (say, an individual and a social layer), the scheme refers to either the influence of the society on the formation of one's ideas (like in prejudice or moral), or the cultural conditioning of one's actions (*e.g.* conscience, or responsibility). This multiplicity of interpretations corresponds to the objective complexity of cognition itself.

*Affects*

Cognition is, in a sense, the primary mental process, since it is closely related to the objective world. On the contrary, in the affective process, $\ldots \to C \Rightarrow S \to C \Rightarrow S \to \ldots$, the active (intentional) component $R$ of the subject is lifted up, and the mediation by outer activity implicitly contained in the link $R \to S$ is entirely hidden, which makes this mental process even deeper immersed in the subject. In the affective process, the passive aspect of the subject's activity is accentuated, with the operational component $R$ replaced by the inner state $C$. Such a process resembles cognition in that a hierarchical picture of the world $S$ is being built; but here, this hierarchy seems to be spontaneously produced from inside the subject, rather than through the subject's observable behavior. In particular, the subject's awareness of an emotion seems to be a result of some inner processes (self-reflection), rather than an outer activity.[75]

However, affective processes imply virtual outer activity, and the indirect character of this dependence only means that the development of a rich emotionality lags behind the development of cognition, and conscious emotions form much later than conscious knowledge. On the other hand, this indicates that the affective sphere is relatively independent of cognition, and the relations between them require a special analysis. Basically, one can distinguish "wise" feelings based on knowledge and understanding from "indicative" emotions, presentiment. Of course, this distinction is relative to the level of hierarchy concerned.

On the lowest level, where the structures $S$, $C$ and $R$ refer to the distinct physiological mechanisms, the mental act $C \Rightarrow S$ appears as dependence of human sensations on the organic processes. In a somewhat refined form, the same idea says that one's instincts determine one's life; the absolutization of the relative independence of affective processes pictures the human psychology as a kind of inner dialog between consciousness and the subconscious (S. Freud). In the same time, psychoanalysis has indicated the only possible way of controlling this mental loop: it is necessary to unfold the link $C \Rightarrow S$ into a complete action $C \to R \to S$, thus restoring the underlying activity, and possibly replacing it with another.

Since the ancient times, emotions were often related to attitudes, and the common description of emotion divided them into "positive" and "negative". This function was said to be in the core of any emotion at all, and all the human feelings were stretched under this dichotomy; the intellectual emotions (like inspiration or curiosity) did not fit into the scheme, and hence they were declared to entirely belong to cognition. In the scheme of affective process $C \Rightarrow S \to C$, all kinds of affects are described, and the discriminative attitudes are placed elsewhere, namely, in cognition. Indeed, the very distinction of the inner structures $S$ and $R$ is already a dichotomy, which is related to the complementarity of the passive and active aspects of the subject's interaction with the world. This means that complex intellectual feelings have much more to do with affective processes than simple binary evaluations, in contrast to the traditional intuitive picture. Such an inversion

---

[75] Among others, the psychophysical theory of emotions by James and Lange was based on this illusion.

of objective relation in the subject is a natural property of reflection, and it is only in conscious, activity mediated reflection that the correct order of things is restored.

Similarly, the distinction of object oriented and general emotions is due to the admixture of the cognitive component. Affective processes are indifferent to the object on themselves; but, since they can only develop on the basis of a certain productive activity, they are always projected onto some reflection of reality. When this prototype is a product of an inner activity rather than external thing, the feeling becomes vague and apparently objectless.

As soon as one goes beyond the primitive organic reactions, the weight of mental feelings will rapidly increase, as compared to all kinds of adaptive self-sensation. In a well-developed subject, each external act is mediated by an inner process, and the affective processes constitute its important aspect. From the viewpoint of biological adaptation, long lasting sentiments are excessive; they only interfere with adaptive reactions and decrease the individual's survival threshold. One can easily observe that animals do not normally have prolonged emotional states, and their feelings are situation driven and transient. In animals, a stagnant mood is an indication of illness. On the contrary, a human without deep feelings is defective, inferior, and underdeveloped.

The scheme $C \Rightarrow S \to C$ also covers such an important class of affective processes as social affects. When the components $S$ and $C$ can belong to a collective rather than individual subject, one obtains the framework for the description of various mass emotions and moods, as well as the affective interaction of an individual with the society. Social psychology thus becomes a full-fledged area of psychological research, dealing with the collective analogs of all the psychological phenomena that are known in individuals.

*Volition*

Like sentiment, the conative process

$$\ldots \to R \Rightarrow C \to R \Rightarrow C \to \ldots$$

does not contain the link $R \to S$, and hence any direct relation to outer behavior. However, while affective processes stress the passive side of the subject, volition is an abstraction of the active aspect, the ability to change the world. In the folded form, this ability presents itself as self-control, complete determination of one's inner state by one's intentions. In this context, without direct attribution to real activity, the transition $R \Rightarrow C$ seems to spontaneously modify the physical states by mere desire to do something. This illusion feeds the numerous self-regulation (or magic) practices based on the idea that once you have imagined some change, you automatically cause it by the very act of mental concentration. The reverse side of the same illusion is the impression of arbitrariness, as if one's actions were only determined by one's inner impulses. In reality, both abilities are limited by the underlying processes of material production and communication. This circumstance was subjectively interpreted in the same volitional terms as an abstract "superior" will that can intervene with one's plans and even entirely destroy them. In popular beliefs, this "higher" will was known under the names of "destiny", "fate", or "god".

Nevertheless, this vulgar picture can be stripped of all the mystic elements and comprehended within the general scheme of conscious activity. Due to the subject's reflexivity, each outer action is normally preceded by a complex inner motion, and conative processes are its indispensable part. All the three classes of mental processes are involved in inner activity, influencing each other or, rather, being the three aspects of the same. While affects are often confused with cognition, will is usually mixed with emotions, and even identified with them. It seems like one *wants* to do something, or *is driven* to do it. This confusion is explained by the absence of the $S$ component in the scheme of a conative process, so that it is in no way *presented* to the subject. That is why it can only be indirectly reflected through cognition and feeling, as a characteristic tint.

The subject obtains the idea of will through lifting the cycle $\ldots \to R \Rightarrow C \to R \Rightarrow C \to \ldots$ into some higher level image structure $S$, which can be schematically drawn as

$$\frac{S}{R \Rightarrow C \to R} \quad \text{or} \quad \frac{S}{C \to R \Rightarrow C}$$

In the former case, the stress is on one's preparation to acting in a particular way, which corresponds to the aspect of will that could be called *determination*. The latter scheme represents the development of the subject's inner readiness for action, *resolution*. Both aspects are present in every volitional act.

The active will has long remained the worst understood aspect of people's inner life, as its systematic study was hindered by ideological premises. The religious idea of the godly will as the source of any activity resulted in moral objection to any scientific investigation, since science had nothing to do with what was attributed to gods. As a reaction, some philosophies advocated unlimited freedom of will, admitting no gods to influence it (Nitzsche). However, despite of all the demonstrative opposition to religions, apology of the human will does not much differ from them. Productive activity as the source and purpose of any mental development was equally overlooked by the both extremities. As a result, the origin of will remained unclear, and it was attributed either to mystical influences or sheer instincts. The understanding of the subject as universal mediation gives the clue to the reflective nature of will as a projection of objective necessity into the subject. The organization of the society suggests the available modes of action, which become the inner states on the subject in the conative process. The availability of all the possible behavioral modes is called *freedom*. Of course, no real society can provide absolute freedom to an individual. The current level of economic development limits the scope of possible operations; this deficiency is reflected in various social limitations, and finally, in the individual's feeling of stress. However, provided the economic conditions are adequately incorporated in the subject's inner activity, one's intentions do not contradict to the cultural environment, and one can remain free even in a non-free society. This is how infinity can exist through final things; what is potential in the outer world is potential in any part. For the subject, freedom therefore appears as comprehended necessity.

Social will is a special kind of will that refers to a collective rather than individual subject. In social psychology, it is well known that a group can behave like a single body, as if it were an individual subject. Unlike chaotic mass motion (such as panic, or economic migration), group behavior phenomena (*e.g.* xenophobia, fashions) are characterized by a significant uniformity of intentions within the group. This uniformity reflects the existence of a particular socioeconomic position occupied by the group, and the stability of the collective subject depends on the conservation of social and economic distinctions. However, in some cases, groups can continue to exist after their economic necessity has long since faded. This is yet another manifestation of the relative independence of mental processes from the outer activity.

In Marxism, the notion of class will was introduced to describe the objective unity of actions within the class as opposed to another class in class struggle. Despite all the differences in individual intentions, the general bias towards denial of the other class' values, and imposing a different kind of values, makes individuals representatives of the common will, rather than mere seekers for personal advantage. This situation is different from mass motion, like panic, when everybody is acting with no regard of the others, in an animal way. The existence of class will is closely related to the development of class consciousness. This is what allows application of the schemes of mental processes to a class subject.

The same schemes can also describe various interlevel processes. On one hand, one finds the examples of prejudice, superstition, religious beliefs, on the other hand, this mechanism underlies leadership, selflessness, or heroism. Thus, if $R$ denotes social action, while $C$ stands for personal moods, the resulting conative process corresponds to tuning one's attitudes to the demands of the common deed; conversely, individual actions $R$ can be adjusted to public sentiments. In more complex cases, several levels of the subject can be combined in a single mental process, which results in a peculiar interplay of individual, collective and global interests in the same activity.

### *From awareness to consciousness*

Though consciousness seems to be related to inner activity, it cannot be associated with neither the primary nor secondary cycle. Thus, the primary process $S \to C \to R$ occurs *before* any outer action, while the secondary act $S \Rightarrow R$ contains folded action, logically *following* it. In this respect, the primary processes are characterized as *sub-conscious*, while the secondary links could be called *super-conscious*; the both cycles remain unconscious. It is only through some combination of primary and secondary levels that one could arrive to consciousness proper.

For instance, the presence of both the primary link $R \to S$ and the secondary link $S \Rightarrow R$ in the scheme of a cognitive process, $S \Rightarrow R \to S$, indicates that a special quality of the internal image $S$ produced: the image incorporates both the properties of the outer world and the reflection of the subject's actions, cognition. Such representedness of the subject's actions in the inner structures produced by those actions is commonly known as *awareness*.

The classical triad of mental processes gives the three types of inner activity the subject can be aware of. One "knows" about one's reflection of the world, emotions and intentions. In their unity, they give the subjective idea of the self.

The primary components of awareness represent the systemic features of the subject, its objective "implementation"; for instance, the link S → C can refer to the physiology of transforming the sensory image into an integral perception. The secondary components of awareness involve societal mediation, thus containing a projection of the culture. In other words, the phenomenon of awareness arises when the physical and biological processes get correlated by an external activity, including both productive actions and communication.

In a way, awareness can be called pre-consciousness, and its primitive forms may be observed in higher animals living with people, when the conscious environment modifies the animal's behavior, synchronizing its physiological processes in a specific way. True consciousness implies synthesis of both primary and secondary paths in the same mental act—for instance, the schemes S → C → R (sub-conscious) and S ⇒ R (super-conscious) should be treated together, as occurring simultaneously. Such a scheme cannot be unfolded into a one-dimensional cycle, being essentially two-dimensional:

On a higher, more socially saturated level, one obtains a similar scheme

which represents the next necessary aspect of consciousness, the feeling of *identity*. In this scheme, the subject's inner state lifts up the outer activity, during which the subjects remains the same despite possible objective and subjective changes.

Finally, the scheme of *responsibility* describes the one's perception of the world in terms of one's own actions, and the subject can clearly see things as the results of activity:

The unity of awareness, identity and responsibility, on the sub-conscious, conscious and super-conscious levels is expressed by the tetrad of inner activity:

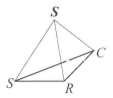

This scheme is a compact expression of all the partial schemes considered in this section, as well as other possible positions of the hierarchy of the subject.

*Ideation*

According to the general principle of hierarchical logic, every link $x \to y$ between two entities necessarily becomes mediated by yet another entity, $x \to m \to y$, and the entirety of such mediations is a higher level entity representing the link itself. Applying this logic to the imagination link $R \to S$, one obtains that there is something (let us denote it as $I$) that represents one's ability to reflect one's own reactions and mediates imagination:

$$R \Rightarrow I \Rightarrow S$$

This something is a part of the subject, since it has been derived from subjective entities; however, it must have a kind of objective existence, occupying the position of an object mediating communication between the subjects:

$$\ldots \to (S \to C \to R) \to O \to (S' \to C' \to R') \to \ldots$$

In its objective form, $I$ must be implemented in the outer things, outside the human body—even outside the non-organic body of the subject. Consequently, a part of the subject's environment is going to lose its "pure" objectivity and become subjective. The process of the subject becoming represented in the outer world, as well as its result, is called *ideation*. In productive activity the hierarchy of the subject grows, and every particular ideation also becomes a hierarchy. An element of this hierarchy (and the hierarchy it represents) is called an *idea*.

To stress the fact that the process of ideation is different from productive activity, a different type of arrows is used in the scheme $R \Rightarrow I \Rightarrow S$. This notation also indicates that the links $R \Rightarrow I$ and $I \Rightarrow S$ are more like mental acts than physical production or consumption. This type of mediation will be called *ideal*, in contrast to *material* mediation in productive activity. Of course, this terminology should be used with reserve, since "material" mediation can be effectuated by such apparently immaterial things as social tensions, personal opinions, or individual skills. Both ideal and material mediation of the subject's self-reflection inevitably combine material and ideal components, and it is only the dominance of one or another aspect, the specific unfolding of the hierarchy, that is meant.

The possibility of ideation is contained in the very opposition of the object and the subject, their mutual reflectivity. An object becomes object only when it is perceived by the subject, and hence is shaped by the subject's intentions. Due to the universal nature of subjective mediation, all the things in the world must become objects, but this universal objectivity unfolds itself differently for different subjects. The subjective aspect of such specific positions of hierarchy is conveyed by the category of ideation.

It is important that ideas are both subjective and objective, representing the subject in the rest of the world. As an object, an idea comes to the subject from the outside, is if it were *given* to the subject by somebody else. As a subject, an idea can mediate relations between objects. Indeed, consider the cycle of subject-object reproduction

$$... S \to O \to S' \to O' \to S'' ...$$

When an object $O$ occupies the position between $S$ and $S'$, two opposite sides of $O$ can be distinguished, namely, its relatedness to $S$ and $S'$ respectively:

$$... S(O) \to O(S) \Rightarrow O(S') \to S'(O) ...$$

These opposites are integrated within the object through a mediating object $I$:

$$O(S) \to I \to O(S')$$

Here, ideation $I$ occupies the position of the subject and hence can be interpreted as a higher-level subject, joining the different aspects of the same object together. This is yet another manifestation of the subject's definition as universal mediation.[76] Folding the mediations $S(O) \to O(S) \to I$ and $I \to O(S') \to S'(O)$ and recollecting that $S(O) = R$ and $S'(O) = S'$, one comes to the scheme of ideal mediation: $R \Rightarrow I \Rightarrow S'$.

---

[76] As a by-product, ideation $I$ gets interpreted as the way to universal internal integrity of the world as achieved through conscious activity.

Ideation explicates the qualitative difference of the link $R \to S$ from the primary links $S \to C$ and $C \to R$, which are not ideally mediated. Ideas exist as a part of a specific culture; they do not belong to an individual.[77] Nevertheless, they are individual ideas, representing quite definite types of subjectivity. The relations between conscious individuals are hence represented as exchange of ideas, rather than products, thus becoming social relations proper.

The cycle of ideation $\ldots I \Rightarrow S \Rightarrow R \Rightarrow I \ldots$ resembles the cycle of mental acts $\ldots C \Rightarrow S \Rightarrow R \Rightarrow C \ldots$, with the social links $I \Rightarrow S$ and $R \Rightarrow I$ replacing the individual links $C \Rightarrow S$ and $R \Rightarrow C$ respectively. This parallelism indicates that ideations can be interpreted as a kind of inner states for some social subject. Obviously, this is the state of a higher level subject, and therefore ideation describes interlevel relations in the hierarchy of consciousness, either embedding one subject into another, or conversely, individualizing a group subject. Reverting the cycle, one obtains the scheme $\ldots S \to I \to R \to S \ldots$ describing the influence of the society on individual reactions. That is, in the subject, the lower level mechanism of complex reflex $S \to C \to R$ becomes complemented with an essentially social mechanism of ideation mediated reaction $S \to I \to R$. The "inner" and "outer" mechanisms work in parallel in every conscious act, which produces a kind of correspondence between $C$ and $I$.[78]

The seemingly subjective ideational mediation encapsulates some outer activities. Such "hidden" mediations can be easily restored, using the schemes for activity mediated by instruments and tools:

$$S \to P_t \to P \to P_i \to S,$$

where the tool $P_t$ and instrument $P_i$ contain both the objective and subjective components, the natural properties and the modes of their usage by the subject:

$$S \to (o_t \leftrightarrow s_t) \to P \to (o_i \leftrightarrow s_i) \to S,$$

or, in another possible position,

$$S \to o_t \to (s_t \to P \to s_i) \to o_i \to S,$$

---

[77] Ideation detaches some aspect of the subject placing it in the outer world as a separate entity The pagan legends of the soul "separated" from the body, wandering spirits, ghosts *etc* is a primitive (syncretic) reflection of ideation—and a sort of ideation too; complemented with an institutionalized system of spiritual oppression (the church), it becomes a religious dogma.

[78] In particular, this mechanism is responsible for modification of animal behavior by social environment.

which can be rewritten as
$$S \to o_t \to I \to o_i \to S.$$
That is, ideation is nothing but the hierarchy of one's abilities and habits, the individual modes of operation with instruments and tools provided by the material culture. These modes are an element of the corresponding spiritual culture, and hence they exist in the whole of material production, outside the subject's body. An ideation contains a product, since habits or abilities cannot exist without application, and their separation from any activity is merely an abstraction. However, the product is hidden in the ideation; it is "wrapped" by subjective attitudes and hence represented in its subjective quality. This complements the treatment of any product as an object, which is necessary to cyclically reproduce the subject-object interaction in any activity:
$$\ldots \to S \to (P = O) \to S' \to (P' = O') \to S'' \to \ldots$$
The understanding of the product as a synthesis of the object and the subject is thus explicated.

On the higher levels of subjectivity, the subjective idea of activity may become associated with ideation as an abstraction of the world and the scheme $O \to S \to O'$ transforms into $I \to S \to I'$, which looks as if the activity were nothing but evolution of an *idea* mediated by the subject. In this abstraction, relatedness to the subject clearly felt in the ideation $I$ can be treated in two complementary ways: either ideas are considered as a part of the subject, and hence no outer activity is possible at all (the solution suggested by *subjective idealism*), or one can believe in ideas existing on themselves, with the subject only mediating their development (the approach of *objective idealism*). The both varieties of idealism do not account for the subject's being a part of the physical world, and the subject's ability to re-create it in reality, rather than mere fantasy. The idealistic illusions are due to the highly indirect nature of most subjective mediations, and first of all, the possibility of using the others as one's instruments or tools. The so called civilized society that has replaced the early tribal system is based on the division of labor and appropriation of one person's labor by another; those who do not produce their living themselves merely collecting the products of the others, are apt to believe that it is their ideas that make the world change, and not the real actions of real men and women (or possibly some other conscious beings).

Language as the universal mediator of communication was often considered as the only possible carrier of ideation, and only verbally expressible ideas were taken into account. However, any ideation is hierarchical, containing both verbal and non-verbal components. One can name anything; but that anything must first exist, it must first shape itself in the culture as clearly recognizable entity (ideation), to become labeled with a commonly recognizable word. Language represents, therefore, the stable core of ideation, its cultural determination. Any creative transformation of the world implies action first of all, and then the words follow.

This does not contradict to the common observation that ideas can drive people to certain acts. Indeed, before an idea can form, one has to act, and one's ideas are products of that preparatory activity, along with other products that arrange the cultural environment in a way allowing the others to "follow" one's ideas. No ideas can be developed by those who don't act and behave. Ideation is a cultural representation of an already existing activity, its public expression rather than cause. It is only in superficial reflection that the cause and the effect can change their places.

The subjective side of any activity thus joins two complementary spheres: inner activity represented by mental processes and outer activity represented by ideation. Conscious actions are always on the boundary between these two areas, requiring both of them. Neither mental nor cultural processes are not directly represented in a conscious act; they are not immediately given to the subject. This is why they are generally called *unconscious*. However, the unconscious is not uniform, being a unity of *the subconscious* (common operations folded in mental processes) and *the superconscious* (cultural predispositions and conditional preferences). While the subconscious encapsulates the subject's *past*, the already assimilated cultural achievements, the superconscious refers to the subject *future*, the modes of behavior that are still to come to awareness and to become conscious actions, and that will then be folded into subconscious structures. That is why the superconscious is said to determine *the zone of imminent development* (L. Vygotsky) for the subject, the range of possibilities for further growth of subjectivity. The simple scheme of the material side of activity

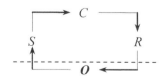

picturing a chain of physical transformations of one object into another, becomes, for a conscious subject, a rather complicated scheme of ideal motion:

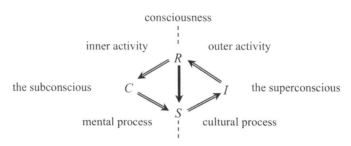

Though this motion occurs inside the subject, it cannot be associated with any particular inner structure, and, in particular, it cannot be confined to the biological body of an individual. Like any collective effect, subjectivity requires a coherent motion of many material bodies, and neither of them is enough for consciousness.

The main role of ideas is to organize conscious behavior concentrating individual activities in the socially important areas. The outer existence of the subject as a hierarchy of ideas makes it susceptible to social influences, since the activity of the others can change the structure of one's ideations and thus induce a change in individual attitudes. Subjectively it looks like a sudden turn in the stream of thoughts, an insight. Depending on the ideation structure, some people can become more sensitive to the social trends that the others; in the most eminent cases, we speak of a genius. However, the very possibility of geniality is due to the cultural dependence of ideation, its formation according to the needs of social development. When the society needs genius, it will give birth to genius.

The scheme of ideation mediated communication
$$\ldots \to (S' \Rightarrow I_1 \Rightarrow S') \to P_1 \to (S \Rightarrow I \Rightarrow S) \to P_2 \to (S' \Rightarrow I_2 \Rightarrow S') \to \ldots$$
can be folded into
$$\ldots \to S' \Rightarrow (I_1 \Rightarrow I \Rightarrow I_2) \Rightarrow S' \to \ldots,$$
which is easily associated with the inner development of one's ideations. Communicating with other subjects, the subject combines their ideations in an individualized way, constructing a unique hierarchy of ideas. On the other hand, the actual existence of subject $S$ is no longer needed in the folded scheme, provided the ideation $I$ is somehow reproduced.

Since ideations are realized as cultural processes, their components (ideas) can, under certain social conditions, become relatively independent of the individual. Different individuals can reflect the same cultural processes and build them into their own ideations. Thus ideas become shared by many people, and eventually form a part of the culture. This shared content of individual ideations is, however, different from material culture, an ensemble of things produced for definite use. This ideal side of culture is called *spirituality*.

An ideation could be considered as individualized spirituality, and spirituality exists through numerous ideations, like any whole exists through its parts. However, like a whole can never be reduced to its parts, no collection of ideations can represent spirituality as such. And, like development of a part results in development of the whole, individual ideations become shared and thus extend the domain of spirituality.

Now, the development of ideas as self-contained entities, $I_1 \Rightarrow I \Rightarrow I_2$ can be understood as the lift-up of individual ideations, with the formation of shared ideas. As soon as an ideation has become an element of spirituality, it can continue to exist even after the original combination of material bodies actualizing the corresponding individual has long since decayed. With conscious beings, the disappearance of a particular implementation does not mean the death of the subject, who continues to exist through other subjects, incorporated in their ideations. Since shared ideas can be reproduced in numerous material forms, they do not depend on the biological life and death, regulating the activity of many individuals for centuries, which adds to the illusion of the ideas existing prior to material things. Once formed, ideas never die; they only develop and show their different facets. Subjects, unlike material bodies and living creatures, are virtually eternal.

This, however, does not mean that any particular individual is bound to live forever. People are not "pure" subjects; their organic bodies can only *represent* some individualized subjectivity. The majority of one's actions is directed to self-reproduction and adapting to the changing environment; such behavior (conscious or not) is not too different from animal existence, and it does not leave any lasting traces in the world. Only the universal content of one's actions is truly *subjective*, and it is this universal core that is kept in millennia.

For collective subjects, ideations are built of shared ideas, and such ideation is essentially an inner process in spirituality. Such processes are different from spirituality itself, even for the humanity as a whole taken as a subject. Collective ideations are the specific positions of the hierarchy of spirituality, and each such unfolding *represents* the hierarchy, never coinciding with it.

The hierarchy of an individual ideation mirrors the hierarchy of the underlying activity. Different kinds of ideas are distinguished by the types of activity organization. They are always ideas *of something* and never ideas on themselves, in no respect to any material things. The most abstract ideas are necessarily implemented in appropriate material formations, at least on the level of expression. In most cases, however, ideas require complex coordination of many activities, rather than mere verbalization, and the formation of an idea goes far beyond simply naming it. Sometimes, the name comes before the idea can form, and this means that there is a clear cultural tendency, which can be felt, but does not yet have found a practical implementation. Much more often, an idea exists without any name at all until somebody happens to put it on the topmost level of an ideation and thus socially represent it.

In general, any idea first appears in the syncretic form, inside a hierarchy of activities and the corresponding ideations; after the activities become common enough, the idea can be abstracted from the underlying activities and *ideally* reproduced in a special activity (social self-reflection); finally, this analytical existence of idea is resolved in a new ideation, compiling all the analytical aspects of the idea into a synthetic whole. The cultural representation of these three stages (or levels of hierarchy) is provided by the levels of spirituality, the ideal aspect of the culture. Thus, the syncretic level of spirituality is represented by both dogmatic forms (beliefs, prejudice, superstition, religion *etc*) and their dialectical complements (e. g. skepticism, nihilism, intuition or fantasy). The analytical level of spirituality includes the three principal forms: art, science and philosophy, which complement each other without being reducible to neither of them. On the synthetic level of spirituality, which is known as *ideology*, one finds such ideological forms as tradition, originality, method, or conviction.

## Hierarchy of Consciousness

Everything in the world is hierarchical. Consciousness and subjectivity are no exception. Because of the universality of subjective mediation, the hierarchy of consciousness is to eventually reproduce the hierarchy of the world. Many hierarchical structures can be distinguished within consciousness and related to the corresponding aspects of the world. In many respects, such partial structures could be treated as relatively independent, remaining the aspects of the same hierarchy. Moreover, individual positions of a hierarchy form a hierarchy too, and this hierarchy admits multiple conversions; every position of this hierarchy represents an individual philosophy as a specific instantiation of philosophy in general.

## *Philosophical categories*

To provide a general context for philosophy of consciousness, a number of standard schemes are presented in this section for reference, without detailed description and justification. Such schemes are implicitly used by any philosopher at all; only few of them take effort to explicate their logic.[79] This book is not intended to develop a hierarchy of philosophical categories; the assorted triads are compiled here in an unsorted manner.

To start with, the fundamental logical triad demands that every phenomenon (a thing) should be taken in three mutually reflective aspects:

*individuality → particularity → generality*.

That is, the thing is first treated on itself, as isolated and self-contained; then we observe the variety of similar things and look at any individual as a carrier of some common feature; finally, one comes to the comprehension of the universal aspects, the idea of the thing. Obviously, this triad merely reformulates the 3U principle of the integrity of the world (uniqueness, universality, unity) in logical terms. The world in general assumes a hierarchy of "worldliness", as well as the infinity of individual worlds.

The triad of the universal aspects of the world,

*matter → reflection → substance*,

has already been discussed in connection to the ontology of consciousness. On the level of particularity, it becomes the hierarchy of the basic aspects of each finite formation:

*materiality → ideality → reality*.

Everything in the world has its material side, being somehow related to matter. In the same way, everything becomes ideal when considered as a

---

[79] Aristotle and Hegel are always mentioned in that respect as the creators of the two fundamental philosophical systems; in between, the formal exercises of medieval scholastics much contributed to the very idea of explicit reasoning.

kind of reflection. However, real things combine both the material and the ideal, thus representing their relatedness to the world as substance.

The same triad taken in the singular sense distinguishes the three aspects of an individual thing:

$$material \to form \to content.$$

Roughly, the material of a thing is what this thing is made of; the form of the thing refers to how its material is organized to produce that very thing; the content of the thing is the unity of material and form determined by the thing's place in the world (the sense of its being, as Aristotle used to say).[80]

Substance in general is an aspect of the world. Due to the self-conformity of hierarchies, a part of the world can accept the definitions of the world in general; this leads to a special triad

$$essence \to appearance \to actuality$$

representing matter, reflection and substance as the attributes of a particular thing (projections); the universal triad is viewed here from the angle of commonality and distinction.

Essence of anything is the unity of its materiality, ideality and reality; it is the expression of the thing's being *in* the world. On the contrary, the category of appearance reflects the particular way of unfolding the essence into something *for* the world (a phenomenon). In the philosophical context, the categories of essence and appearance reflect the ontogeny and phenomenology of the thing, respectively. The synthesis of essence and appearance, actuality indicates that no appearance is possible which would not be implied by the thing's essence, and nothing in the essence is hidden from appearance; all the latent features have to become actual.

The essence of a thing refers to the possible manifestations, *potential* existence.[81] The *actual* existence is a definite entity, as distinct from similar and different entities. However, individual things can never represent their essence in full, just because of their uniqueness. It is only through different appearances that the essence of an actual thing can be revealed as the common core of different phenomena. Conversely, an appearance is needed to transform the thing's essence into actuality, developing an abstraction to

---

[80] This triad has been close to explication in esthetics. According to many critics, true art demands the unity of material and form, as well as the harmony of expression and content. Somehow, philosophers did not notice that these two oppositions form a triad.

[81] The possibility of existence is a kind of existence itself. If something is possible in principle, it is already present in the world as possibility.

something concrete.

From the systemic angle, the actuality of a thing implies transformation of essence to appearance, and back. Hierarchical understanding of actuality makes essence and appearance mutually reflected and developing through each other.

## *Reflection schemes*

The hierarchy of reflection is generally represented by the triad

$$existence \to life \to activity$$

that has earlier been presented to indicate the place of consciousness in the world. The universal levels of existence

$$being \to motion \to development$$

correspond to the levels of organization:

$$structure \to system \to hierarchy$$

Similar schemes for life and activity could also be established; they will be introduced elsewhere. There are many other dimensions in existence life and activity. Thus, anything that exists is either a thing or its aspect, which gives the triad

$$thing \to aspect \to entity$$

On the level of life the same triad takes the form

$$body \to soul \to organism$$

with any organism treated as the unity of the body and the soul. Finally, on the level of activity we obtain the already discussed triad

$$nature \to spirit \to culture$$

indicating, in particular, that the soul is the necessary premise of spirit and its animal analogue. While the aspects of a thing refer to its external definiteness and can be multiple, the soul refers to the internal definiteness of the organism and is unique; the spirit is relatively independent of nature, uniting both external and internal determination.

The latter scheme is the world level form of the singularity level triad

$$object \to subject \to product,$$

which describes the most universal hierarchy of activity. The world as an object is called nature; the category of spirit is to express of the world's universal reflectivity, while the category of culture pictures the world as a product. In every instance of conscious reflection, the world is duplicated as both nature (existing before the subject) and culture (the "second nature" created by the subject to respond to the subject's demands).[82]

As usual, the general triad *nature → spirit → culture* can also be treated in the attributive way, distinguishing the natural, spiritual and cultural aspects in any object, subject or product. Finally, the components of the triad can be individualized, to consider many different natures, spirits or cultures.

## *Levels of subjectivity*

Considering the spirit as universal mediation between nature and culture (as distinguishing nature from culture), one will distinguish the two sides of spirit relating it to nature and culture respectively. The "natural" aspect of the spirit presents it as "inhabiting" a material body[83], a thing separate from other things; this stresses the subject's individuality. For any natural body, the subject is nothing but another body, albeit behaving in a peculiar way not always consistent with natural laws. On the other side, the "cultural" aspect of the spirit could be called sociality, since it refers to the place of the subject in culture, projecting spiritual culture onto an individual as its mental organization. The link between the two poles could be related to personality. Hence the triad of subjective forms characterizing the modalities of the subject:[84]

*individuality → personality → mentality*

This triad primarily refers to the results of the development of subjectivity. Alternatively, the objective hierarchy of consciousness is expressed in the triad

*awareness → identity → responsibility*

Awareness is the most folded form of consciousness. It is rigid enough to

---

[82] It is in culture that human creativity gets virtually embodied, thus giving people eternal existence. Culture is a human-made world, and human can feel themselves as demiurges in respect to this new world, which becomes syncretically reflected in the idea of a god.

[83] This does not need to be a single human body, or even a number of humans as representatives of a biological species—within a definite culture, a non-organic body can serve for the "embodiment" of Spirit as well. Traditionally, a single person is the most prompt association for "an individual"; however, the idea of a group of people as an individual has been extensively discussed in the literature as well.

[84] This conforms to the classical division of psychology into general psychology, differential psychology (studying personal traits) and social psychology.

appear already on the organic level; its functioning can be significantly depending on the organic implementation. Identity refers to the presence of the Self. This is the central level of consciousness playing an integrative role. It is closely related to communication and language, being the least possible in animals. Responsibility relates consciousness to activity (motivation), and to the active rearrangement of the world by the subject. This is primarily the social side of consciousness (and hence inherent to any kinds of consciousness at all, since consciousness in general is socially mediated). The traces of responsibility can be observed in animals when the animals live with humans for a long time, sharing the same social hierarchy.[85]

The general direction of development is indicated by the triad

*consciousness → self-consciousness → reason*

While consciousness proper is directed to the recognition and assimilation of outer objects (possibly incorporating lifted-up inner formations), self-consciousness forms in reflective activity specially organized to produce the subjects rather than outer products. Reason is the synthesis of consciousness and self-consciousness; it is based on conscious creativity, which employs natural directions of development to put this development under conscious control.

## *Levels of culture*

As anything in the world, culture has its material and ideal aspects, which could be formally expressed by the triad

*material culture → cultural reflection → cultural formation*

Each cultural formation is thus understood as self-developing substance, the unity of "cultural matter" (all the products of conscious activity as a trace left by the subject in the world) and "ideal reflection" (the same collection of products viewed from their ideal side, as the modes of activity). Cultural formations can be established in quite different respects; commonly, cultures are distinguished by geographical or ethnical criteria, as well as by the historical epoch. The development of humanity in general goes through a sequence of cultural formations, and each formation is characterized by the way of material production (socio-economic formation) and the way of

---

[85] This is different from the natural behavior of an animal in a hierarchically structured community.

ideation (related to the type of people's involvement in cultural processes).[86] Correspondingly, cultural reflection in a cultural formation is characterized by the type of material production and the type of ideation.

Considering cultural reflection, one can return to the universal formula of activity, $O \to S \to P$, and distinguish the passive (objective) and active (productive) aspects of reflection, transformed to each other by the cultural projection of subjectivity as such:

$$cultural\ experience \to spirituality \to praxis$$ [87]

The mode of life and the usual ways of doing anything are reflected in the category of cultural experience. This is how people live and what they do. On the other hand, praxis refers to historical development, reorganization of people's life, and hence their cultural experience. Spirituality determines the direction of cultural development, relating it to conscious activity. This is the link between cultural experience and praxis, the way praxis differs from mere living.

*Levels of spirituality*

The development of spirituality follows the general rules of hierarchical development, as reflected in the principles of diathetical logic. Since spirituality grows from experience into praxis, it must manifest two aspects reflecting the two opposites, as well as an intermediate level, that mediates the connection the experiential and practical aspects. Thus one comes to the hierarchy of spirituality:

$$tradition \to creativity \to cultivation$$

In this scheme, tradition is the syncretic aspect of spirituality, with the agent of activity yet identified with that activity, and the subject merged with the object; cultural reflection leads to the reproduction of the already existing forms. This is the earliest form of spirituality providing the necessary background for the higher forms. On the level of creativity, reflection becomes analytical: the subject is consciously opposed to the object, and the form of reflection is essentially different from what is reflected. On the

---

[86] The category of socio-economic formation has been introduced by K. Marx; however, the complementary reflective aspect of cultural formation was overlooked by Marxism, though all the relevant ideas were already contained in the manuscripts of Marx, Engels and Lenin.

[87] In a way, the Aristotle's triad of the basic activities is thus reformulated: theoria (contemplation), poiesis (creativity) and praxis (activity on itself).

highest level, the integrity of the product is restored; creativity is directed to extended reproduction, while reproduction is organized so that it demands wider creativity.[88] As usual, the categories in this scheme can be considered as separate entities, as the mutually reflected aspects of something, or as the levels of some inner hierarchy.

Unfolding the mediating category according to the rules of diathetical logic, one obtains the hierarchy of analytical spirituality:

$$art \to science \to philosophy$$

On the level of tradition, these three components are represented in a syncretic way, merged together; one can rather speak about artistic, scientific or philosophical trends in people's acts and common products. With further development of creativity, art, science and philosophy become separate activities,[89] up to assuming institutionalized forms.[90] In cultivation, different activities become the distinct aspects of the same; art, science and philosophy are fused in universal forms of reflection serving as cultural triggers of the reconstruction of the world.[91]

Since art, science and philosophy are implemented in specific activities, at least as their aspects, they can be reflected as a part of cultural existence and produce various "higher-order" formations in creativity: each activity is viewed from the angle of its inner logic, but the analysis of that activity is also an activity with its own logic *etc.* For instance, art criticism occupies a special niche in the arts, being a kind of "philosophy in art". Similarly, science pays much attention to methodological study, which represents philosophy in science. Also, there are art-like levels in science and philosophy; science too penetrates philosophy and art. Thus the levels of

---

[88] Tradition is essentially empirical; creativity is based on abstraction. Cultivation could be called applied spirituality, the path from abstraction to concreteness. That is, the way of synthesis provided by cultivation is abstraction from abstraction, the idea of *application*. Cultivation is governed by the necessity of finding possible applications for any abstraction, correlating it with reality, yet in an abstract way. When the abstractness of application gets removed, spirituality grows into praxis.

[89] This process has its own historical stages; it could be considered as the step from consciousness to self-consciousness, within spirituality.

[90] Tradition, creativity and cultivation can as well be considered as the aspects of any activity at all. However, art, science and philosophy as the levels of analytical spirituality do not necessarily imply specialization and division of labor. It is only under definite social conditions that the three modes of reflection can become separate occupations, professions. Institutionalized spirituality loses its spiritual character, becoming a part of cultural experience

[91] Art, science and philosophy could also be considered as spiritual analogs of the subconscious, conscious and superconscious levels of activity.

creativity become reflected in each other, and self-reflected.

*Religion and ideology*

Religion belongs to the level of syncretic spirituality. It is based on tradition and maintains tradition; the central form of religious reflection is *dogma*. On the contrary, ideology is a part of cultivation; this is a mechanism of promoting new ideas and testing their viability. While religious consciousness is made of *beliefs*, ideological consciousness grows from *conviction*, a general idea that has been practically supported and proved its efficiency. Convictions are never rigid like religious dogmas; they demand critical reflection and revision in accordance with the present cultural situation and the direction of cultural development.

Religion could be called a primitive (early) form of spirituality. This does not mean that religion is simple—on the contrary, lack of development leads to cumbrous dogmatic systems that are imposed on the people via intricate rites and ceremonies. Ideology is often more straightforward, since it demands definite action immediately related to a fundamental idea; however, ideology is not primitive, implicitly containing the whole history of spiritual development. Regardless of their complicatedness, any religion is primitive in the sense of preceding self-conscious spirituality as such. Religious attitude to reality is characterized by lack of comprehension; it is akin to a childish refusal to deal with things that cannot be grasped at once.

Being a level of spirituality, religion often pretends to be its only possible form. Such pretence is utterly unjustified in most cases, since dogmatic thought lacks freedom, confining the spirit to a collection of uncritically excepted beliefs. Nevertheless, religion is objectively necessary on certain stages of cultural development; it can support spiritual progress, and a belief sometime becomes the first stage of conviction. Institutionalized forms of religion have nothing to do with spirituality at all.

*Art*

As the first break of tradition, art "translates" experience into a spiritual representation, suppressing irrelevant details and refining the universal content of any cultural phenomenon. That is, art is essentially abstraction; it cannot and should not imitate nature. The apparent similarity of realistic art to actual life still assumes that art is still different, and artistic truth has nothing to do with mere imitation. In the arts, the form of activity becomes abstracted from the activity itself, which allows to combine forms in an arbitrary manner and thus obtain yet unknown combinations.[92]

However, the way of abstraction in the arts is still inseparable from experience. There are no specifically artistic activities, since *any* activity at all can become art, as soon as ordinary work goes beyond mere skill. *Perfection* could be considered as the core category of aesthetics; it is what distinguishes art from plain activities. The abstract nature of art is thus explicated: since no common experience can be perfect, activity has to be "refined" to achieve perfection, to reveal a universal core. Such "purified" experience is the elementary construction block of art, an artistic *image*. Complex images can be constructed from simpler images, remaining as syncretic. Images of art can be produced, but not be defined or formalized; they refer to other images in a syncretic way, through imitation and allusion.

Since art is grows from within experience, any person has to discover an individual way to perfection. Nobody can learn, or teach art. Studying the history of arts, technical tricks or traditional patterns (styles) does not make one an artist. Aesthetic education is useful for general development, it enhances creativity, but it cannot suggest any recipes of extracting the eternal from the transient.

*Science*

When a number of abstract ideas have taken shape in the arts, new abstractions can be built on the basis of these primary ideas, without direct reference to experience, as if they were reality themselves, or at least direct representatives of reality. This kind of spirituality is characteristic of science.

Scientific *notions* are more abstract than the images of art, since they are related to reality through many intermediate stages. On the other hand, this makes them more universal, and hence applicable to a wider range of apparently incomparable situations. Notions can be formally combined and formally constructed, being mutually defined through their place in a hierarchical structure (scientific theory). Science detaches the subject from the object, presenting its results in an objectivated form (knowledge), which opens way to universal propagation of scientific ideas, since everybody can be *taught* any science.

Still, scientific knowledge is as far from complete comprehension of the world as artistic imagination. Art and science are two complementary kinds of abstraction, they are equally necessary for spiritual integrity, though this

---

[92] However, art does not invent new forms; it only borrows them from various activities.

integrity can only be achieved through the synthesis of the both.

Like art and philosophy, science is reflectively represented on all the levels of spirituality. Syncretically, it is merged with activity, contained in its historically elaborated schemes, apprehended through learning. With the development of society, science becomes associated with a separate activity appropriated by a special social group, professional scientists. This is what is commonly meant under "science", its analytic level. However, science has to finally become a part of practice, and thus return to the people's everyday life. This level may be called synthetic, and one might suggest engineering as a representative.

*Philosophy*

Scientific ideas derived in a chain of formal conclusions are void unless they can be somehow imagined. On the other hand, an artistic image cannot be comprehended without preliminary training, which associates one image to another thus making it similar to a scientific notion. In this way, art and science penetrate each other; their unity forms the next level of spirituality, philosophy.

Artistic and scientific types of abstraction are synthesized (and lifted up) in philosophy, which is not yet concrete, but demands concreteness and prepares it.[93] This determines the dual nature of philosophy: primarily, it is the highest level of abstraction introducing the idea of the integrity of the world; however, since there is no more room for abstraction, any further reflection must be concrete. That is why philosophy is readily involved in ideological struggle and practical activity.

The analytical type of creativity is still retained in philosophy, since its product (an individual philosophy) is still different form the areas of the culture reflected it reflects. However, there is no preferable way of expressing philosophical ideas: they can be conveyed through the literature of any kind (a philosophical treatise, scientific papers, or belles-lettres); alternatively, philosophy can propagate through the practical efforts of many people struggling for a common cause.

Philosophy is the most reflexive kind of spirituality, and its hierarchy is virtually identical with the hierarchy of the world. Philosophy influences art and science, as well as its own development. However, this influence is not direct, and philosophy as a regulator of artistic, scientific or philosophical

---

[93] Philosophy could be said to bring the abstractions of art and science back to activity.

creativity appears in the transmuted forms. In relation to art it becomes *aesthetics*. The scientiflic type of creativity is reflected in *logic*. Determining universal self-development, it manifests itself as *ethics*. The structure of philosophy thus immediately reproduces the structure of creativity, and every individual philosophical teaching contains aesthetical, logical and ethical principles. Aesthetics, logic and ethics are often considered as a triad of philosophical disciplines. However, they cannot develop separately, requiring mutual reflection. Formal separation of one aspect of philosophy from another is contrary to the principle of integrity and hence reduced philosophy to science or art.

The basic structural element of philosophy is philosophical *category*, which could be considered as the hierarchical synthesis of artistic image and scientific notion. Categories represent the typical (universal) ways of action, and hence they are not as dependent on perception as the images of art, while being less abstract than the notions of science.

*The universal ideal*

Human spirituality is the highest level of the ideality in general. Any reality in the world is the unity of the material and the ideal sides, and thus the ideal component is necessarily present in everything, though assuming different forms on different levels of reflection. The lowest, existential level of ideality is characterized by all-penetrating syncretism, and that is why most philosophies do not distinguish reality from matter, or reflection. On the level of life, a living thing becomes opposed to all the non-living things, and its ideality becomes analytical. The polarity of inanimate and living things is reflected in the idea of *the soul*.

Logically, there must be a synthetic level (conscious activity), where ideality will be neither opposed to the rest of the world, nor syncretically merged with it. Any individual activity bears universal significance, thus making a single person a representative of the whole world. This universal core of the ideal is *spirituality*.

The universality of the spirit has many important implications. Thus, not any behavior may be called conscious (that is, spiritual), but only such kinds of behavior that are universal in some respect. Consequently, it is unimportant for conscious activity, in which organic (or non-organic) forms it will be implemented. In other words, the spirit is not contained in any body— rather, an individual spirit is an individualized form of the unity of the world.

The natural corollary is that playing with images, notions or categories (which is commonly associated with art, science and philosophy) has nothing to do with spirituality as long as it reflects something specific or individual; true spirituality begins where one's needs express an objective necessity, and one's will represents the unity of many partial wills. The products of art, science and philosophy must carry some universal content, regardless of the elaborateness of their material and form. A professional artist may be very skillful in designing new arrangements of forms—still, only few of these combinations will reflect the universal in human activity, thus becoming the instances of art. Similarly, the skills of a professional scientist do not imply ability to increase knowledge or wisdom. Professional education gives one a collection of tools and instruments, but it cannot make one spiritual.

# EPISTEMOLOGY OF CONSCIOUSNESS

According to the definition of the subject as universal mediation, it is in the nature of the subject, to comprehend themselves as a part of the world. Nothing can avoid being assimilated by the subject in its internal replica of the world, and then being transformed into a product, a part of the world rebuilt by the subject. The subject's consciousness is not an exception: it is both the result of natural development and the product of conscious activity. This active self-construction is one of the distinctive features of subjectivity.

Each thing becomes an object when it is presented to the subject. The very definition of an object implies the subject, and no single thing can be perceived without an admixture of subjectivity. We see the world only through our activity, and in relation to it. Historically, this trivial circumstance was reflected in numerous paradoxes and resulted in the distorted vision of consciousness and a conscious being.

In the most radical forms of subjective idealism, it was declared that, since all we perceive is perceived through our senses, nothing should be considered to exist beyond our sensations, and it would be meaningless to ask about reflection of anything outside the individual subject. This idea led to obvious contradictions and inconsistencies; however, the majority of the adepts of subjective idealism preferred to blindly ignore them. Indeed, if there was a single individual knowing nothing but his sensations, why should one care for any knowledge at all? If there were no other people to communicate with, there would be no need of cognition, and all the only existing individual does not need to perceive anything at all. There would be no way to learn about one's own sensations, since any such knowledge would already oppose them to the subject as something (at least partially) external to it. The only consistent state of such an isolated individual would be a uniform nothing, with no motion at all. This conclusion is experimentally confirmed through observing people in the conditions of sensory deprivation. In such experiments, after a transitory surge of hallucinatory perceptions, the person tends to seize any activity and fall into a kind of lethargy, somnolence without dreams. Numerous meditation practices used that fact to reach a state of indifference they named "nirvana", "enlightenment" *etc.*

Thus, starting from the impossibility of any knowledge but self-knowledge, one comes to impossibility of self-knowledge as well. In subjective idealism, there is no epistemology of consciousness. Any attempts to speak about self-comprehension in subjective idealism are necessarily eclectic; they always employ logically alien elements. Since subjective idealism does not care for knowing anything at all, such eclecticism often pretends to replace the very idea of knowledge, with meaningless babbling put in place of science.

Feeling the utter inadequacy of such an approach in real life, many philosophers tried to disguise subjective idealism, pretending to avoid the very question of objective existence by saying that we just cannot know about it, and therefore should not talk of it at all. This attitude appealed to scientists, whose poor philosophical education did not allow revealing the true face of that school, collectively referred to as *agnosticism*. The normal indifference of a scientist to anything that cannot be scientifically tested was thus substituted by denial of anything beyond the scientific fact, which is nothing but an eclectic variety of subjective idealism merely extending the physiological senses of an individual subject to the instrumental data and formal conclusions constituting the "senses" of the academic community as a collective subject. The agnostic consciousness is not entirely blind, like that of subjective idealism, it is only strongly myopic.

A much more consistent view was put forward by objective idealism. In this branch of philosophy, the whole world is declared to be a product of some supreme subject, differently named by different philosophical schools: the absolute spirit, God, the supreme will, the fate, karma, Tao *etc.* All varieties of objective idealism are essentially about the same: first, the world gets created by the subject, and then it becomes comprehended by it, thus restoring the unity. In this case, cognition is a kind of self-cognition; for objective idealism, the world is definitely comprehensible, and the subject is capable of self-cognition through the subject's own products. In objective idealism, the extension of the individual to the collective subject reaches the ultimate form of considering any subject at all as a manifestation of the absolute, universal subjectivity.

In this philosophy, one can logically admit science, and speak about studying the world. Any partial subject can do research and discover things and other subjects. Individuals can communicate their knowledge to other individuals and establish common conceptual frameworks. Objective idealism perfectly matches the commonly perceptible process of producing things, and learning about things produced by the others. On a certain stage of social development, when people become less dependent of nature than from other people, they are tempted to consider their cultural environment as the only environment one can ever have and all things are deemed to be made by somebody. If there is nobody human to create it, a non-human entity beyond human comprehension can be easily fantasized, which, however, is felt to be somehow related to human activity.

The problem with objective idealism is that there is no way to tell, why that absolute entity creating the world is necessarily a subject. There is nothing subjective about it, it exists regardless of any other subjects and develops according to objective laws. Why should we call it a subject? Why not simply admit that this is the world in general, which develops through numerous partial manifestations, including its manifestation to the conscious subject, nature? Consistent objective idealism is a straight route to materialism, since the very assumption of an objective process of spiritual development admits an object prior to any spirit, and the only logically consistent continuation is to reverse the scheme and start with the object (nature), deriving the subject (spirit) from nature as a result of natural development.

Materialism, in contrast to idealistic philosophies, tried to describe the world as existing regardless of human activity, and those who cope with anything practical will necessarily act as spontaneous materialists to be successful. The most convinced idealists immediately become quite materialistic, when it comes to food and shelter, to health and wealth. A solipsist, who writes the books on that there is nothing in the world but his imagination, will call a real doctor at a slightest uneasiness, and a real policeman to defend himself of a street robber. This most ancient kind of materialism governed the work of many scientists ever since science has separated itself from the arts and philosophy, constituting a relatively independent cultural sphere. The scientist believed that there is an object to study, and the subject, who studied the object, producing a commonly acceptable way of treating objects, knowledge. The main goal of science was called "truth", and the truths (facts) had to be *discovered*, presuming their pre-existence in nature. Once discovered, a fact could not change, becoming a little stone in the huge pyramid of absolute knowledge.

The scientists could easily remain within that primitive philosophy on the early stages of the development of science, when it dealt with relatively simple things. However, as soon as scientific inquiry has reached the domain of very complex motion and development, the traditional scientific objectivity failed to adequately explain phenomena, and the self-confidence of a spontaneous materialist was shattered, when research became indirect and less intuitive. The epistemology of consciousness was the first and heaviest stumbling block for natural scientific materialism.

Since consciousness was thought to be a natural property of an individual, materialists tried to attribute it to some particular organ, or to distribute it between the organs of the human body. The ancient theory of four temperaments attributed modes of human behavior to the proportion of the four fluids: blood, bile, black bile and phlegm. In the beginning of XX century a similar approach became rather popular, taking the forms of the James-Lange theory of emotions, psychological behaviorism *etc.* By the end of the XX century, a philosophical school known under the name of "consciousness science", despite many criticisms, established itself as a standard of scientific methodology in any consciousness studies. In this school, subjectivity was believed to be a function of the brain, and only the study of cerebral processes was recognized as "scientific". Logically, this implied that consciousness should be genetically pre-determined, and the origin of mental disease was sought in bad inheritance. Official psychiatry believed in powerful chemicals as the only cure for psychotic patients, and medicine was often sacrificed to the profits of pharmaceutical companies.

However, scientists felt that such a reduction of the subject to mere physiology did not solve the problem, but rather pushed it out of sight. These doubts took the form of paradoxes (L. M. Vecker):

1. *Ontological paradox*: higher level psychological phenomena cannot be described in terms of lower level mechanisms. In particular, no psychological phenomenon corresponds to a unique physiological pattern, and no physiological process is unambiguously associated to a specific psychological effect.

2. *Epistemological paradox*: people's perceptions and intentions are always expressed in terms of outer objects rather than physiological or psychological characteristics. We observe outer things, rather than our

feelings, and our goals are things outside us. Even reflecting on our own moods and feelings, we do it as if they did not belong to us, and perception of a feeling is different from feeling itself. This was metaphorically described by the idea of a *homunculus* inside each person, the one who observed our conscious actions and reported them to us.

3. *Ethical paradox*: consciously perceiving ourselves, we change ourselves and thus make our perception obsolete. As soon as we have established a law of mental dynamics, we can consciously violate this law through an outer activity specially directed to that change. It seems like, in the science of consciousness, there can be no final truth, or universal laws, and all we can know is limited regularities.

These paradoxes were often used to "prove" the insufficiency of scientific materialism and the inevitability of idealism, at least when consciousness and subjectivity are concerned. However, as it is usual with paradoxes, they are entirely due to an artificially narrowed view, inadequately applied to a wider area, where wider notions should be used.

The ontological paradox is easily solved admitting that reality is always hierarchical, with higher levels providing a general context for lower level processes, and the reverse influence, from lower to higher levels is possible on the average. Each higher level process can be represented by many lower level implementations, neither of them being better than another. Conversely, very different configurations on the lower level can correspond to the same higher level state. The very difference of "the lower" and "the higher" is relative, and depends on the context.

The epistemological paradox is due to the illegal identification of the conscious subject with the physiological body of an individual. As soon as we admit that consciousness is the attribute of both the organic and inorganic body of a person, there is nothing strange in that the inner states of the subject are expressed in terms of outer things. Consciousness does *not* belong to the subject; rather it is a way of the subject's involvement in objective processes. This makes our inner states observable to other people and us. Since any individual is considered as a part of the society, the subject becomes hierarchical; any self-perception is mediated on different levels by the society, and the true "homunculus" can be easily found outside rather than inside us, in the people surrounding us. We see ourselves by the eyes of the others, and nobody else can tell us who we are.

The ethical paradox is trivially resolved considering knowledge as a hierarchy and taking it in development, rather than statically, as huge heap of unchangeable truths. Any truth is relative since it can only exist in a specific cultural context. However, any truth is also absolute in that it will always be true in an appropriate context, and the corresponding piece of knowledge will be applicable every time when certain aspects of that context become culturally reproduced. In particular, due to the hierarchical organization of culture, old truths can reign on some deeper levels of hierarchy, even though the overall behavior has long since evolved to something entirely different. Specifically, when we comprehend ourselves, the hierarchy of the subject grows, and we indeed know ourselves, though only in certain respects. More knowledge comes with time, and there is nothing that cannot be learned.

Therefore, there is no need to appeal to any incomprehensible supernatural entities in studying subjectivity; the materialistic picture of the subject as a part of the self-developing world provides a consistent and uniform platform for any science, including the science of consciousness.

## Hierarchical Methodology

A review of the modern attempts to approach scientific methodology in the study of consciousness leaves one surprised with the inefficiency of the efforts made, despite the heavy attack with all the means available in science from late 1990s well into the next century. So far, no unifying idea can be felt behind the multitude of models and variety of conceptualizations present in the literature; no common platform for discussions and special research. To some extent, this might be due to the versatility of the subject itself; however, the major drawback of modern consciousness studies is the inadequacy of the very logic underlying them.

According to the hierarchical approach, the methodology of any science must reflect the organization of the object of study. As soon as the subject becomes comprehended as universal mediation, the main purpose of science is to analyze the special forms of human activity and indicate their universal content common to the possible partial implementations. Scientific research will demonstrate how universal mediation can be effectuated via certain organic and social bodies, despite their finite and limited existence.

On the other hand, scientific research should not be confused with artistic or philosophical studies, which have their own niche in describing consciousness and subjectivity; similarly, science is not directly related to its practical applications, or to the ordinary life. The motives of a scientist do not coincide with the interests of a novel writer, or a politician. Nevertheless, the hierarchical nature of the subject leaves enough room for diversity even within the scientific angle of view. There can be numerous special sciences of consciousness, but they will necessarily have something common, since they all study the subject and subjectivity. The self-conformity of any hierarchy implies that every special science will follow the same methodological line, to be consistent and adequate. To become a sound basis for scientific research, this methodology is to be derived from the most general principles; this is especially important for studying consciousness, which is universal by nature. One or another scientific position is not a matter of preference; they must obey the same universal logic of research.

## *Between the object and the product*

As any natural thing, the subject reflects the world, being one of its parts. The world is basically represented in any of the innumerable things it consists of. However, such syncretic reflection is not enough to be knowledge. Even the organic assimilation of the world and adaptive behavior is far from conscious knowledge. To become knowledge, the internal image of the world (objectively a part of the subject) must be related to the objective phenomena in a universal way, that is, through another subject. Knowledge is objective, since it adequately reflects nature; however, every object is defined through its relation to the subject and hence contains a subjective component. On the other hand, no object can be reduced to its subjective side; it is primarily a material thing, or a relation between material things, and never an abstract play of imagination.

Ontologically, the synthesis of objectivity and subjectivity can only be achieved in a product of conscious activity. The only objective way to subjectively assimilate anything is to reproduce it as a product—this holds for the subject's self-comprehension as well. All we can know is only the products of our activity; we observe our own trace in the world to understand it and ourselves in it. However, this has nothing to do with agnosticism, since conscious reconstruction of the world is an objective process, and the universality of the subject ensures that there is nothing in the world that could not be involved in the subject's activity.

Consequently, to study consciousness, we must be able to imprint it in our products and thus make observable as an outer thing. The most methodologically important corollary is that subjectivity cannot be directly observed, since the inner structures or processes of the subject would not possess the specifically subjective quality in direct observation, as the triad $O \to S \to P$ indicates, where the position of the subject (mediation) is between the object and the product. The scheme $S \to S'$ cannot represent the reflection of subject $S$ in subject $S'$, since, in this scheme, $S$ occupies the position of an object, while $S'$ occupies the position of the Product, so that there is no mediation at all, and the scheme rather refers to the objective development of the subject, $O \Rightarrow P$. To describe subjective reflection, the image of $S$ in $S'$ must include something produced by $S$, as well as the very process of production (or subjective mediation), which is the only possible manifestation of subjectivity. To represent this, a more elaborate scheme is used:

$$(O \to S \to P) \to S' \to P'$$

Converting the hierarchy, we obtain the scheme

$$O \to (S \to P \to S') \to P',$$

which says that the only way for the Subject to comprehend itself is mediating the very subjective mediation by communication, which is essentially exchange of products. To maintain the universality of subjective mediation, the product must be as universal as the subject, being a kind of exteriorized subject. That is, unlike the outer products which do not function as products outside the cycle of their reproduction, the universal objective mediator of subject-to-subject communication will preserve the traces of subjectivity in a much wider range of situations, being relatively insensitive to the specificity of individual communication acts. Such a universal product, joining the subjects in the universal way into a higher-level integrity is readily identified with *language*.

Issues of language and speech behavior will therefore occupy the central place in the scientific study of consciousness. One can never be quite sure in the validity of any conjecture about subjective phenomena without tracing its consequences for language. However, language is not the only key to comprehending consciousness. Since any product (and virtually any object) contains a subjective component, the descriptions of the material things and processes from the viewpoint of subjective mediation can bring valuable knowledge about the mechanisms of consciousness and its forms. There are two main directions of that study, one uncovering the essential subjectivity of reflecting objective phenomena by the subject, and the other investigating the specificity of the secondary objectivity of the world as modified by the subject. That is, one either considers the influence of the subject on nature, or, conversely, treats the subject as a specific property of the artificial environment it creates. In consciousness studies, the former direction might be called "(psycho)physical" (in a wide sense, including all the varieties of physiological, neurological and behaviorist methods, as well as certain aspects of economic science and sociology), while the latter approach is characteristic of culturological research (including all kinds of history). An adequate description of consciousness must combine the both approaches, indicating the ways they are related to each other.

The universality of the Subject means that studying consciousness cannot be restricted to any single science (and even to science only).[94] There is no dedicated "science of consciousness". Every part of the world is bound to be transformed and assimilated by the Subject; and every part of reality is a source of knowledge about consciousness. Moreover, due to convertibility of hierarchies, the same object can be studied from different angles, including both subject-related and non-subject aspects. For instance, a human being can be studied in quite different ways: as a material body moving according to mechanical laws, or a thermodynamic machine, or a chemical reactor, or a living organism, or a cybernetic device, or a conscious individual, or a social relation, or a role in a group, or a mediator of global processes, or a link between the most distant formations of the Universe. Some of these studies will have something to do with consciousness, while other sciences have their own goals, only providing the necessary environment and background for investigating consciousness. On the other hand, sciences about various aspects of consciousness can serve as a background for some other research (*e.g.* physiology of the brain, or technological development); cultural influence on natural things demands important corrections to their description, since an isolated thing behaves differently from that involved in conscious activity.

Objective reflection is related to the development of the world; subjective reflection, the reproduction of the subject within itself, is the source of the formation of the internal model of the world commonly associated with knowledge. Knowledge is primarily a product; this implies both subjective product and objective product, reproduced in the objective and subjective cycles of activity respectively. The both are necessary for the

---

[94] From the culturological viewpoint, every science (or other form of social activity) represents some aspect of the subject, explicating the corresponding inner formations.

whole; the objective forms of knowledge and its subjective representation are complementary, though they never coincide.[95] In particular, this means that any inner picture of the world we might have is also a reflection of ourselves, that is, our modes of using things and our ability to change them in a desirable direction..

In the subject's self-reflection, any question is essentially reflexive, and it is by the reflexivity of the answer (the product) that the different modes of our comprehending ourselves can be distinguished. In scientific research, this reflexivity is present in its analytical form, in contrast to the syncretic reflexivity of the arts and synthetic reflexivity in philosophy. Science borrows the basic idea of its scope from the syncretic level, while many partial scientific pictures of the world become integrated in a philosophical doctrine. Philosophy of consciousness is interested in the relation of analytical knowledge to the whole; a scientific approach would treat consciousness in a special way, within the scope of the particular science. The same phenomenon will be differently described by different sciences, each of them selecting its own angle of view, so that some features are accepted as relevant, while all the rest is treated as "noise"; the philosophical approach is to indicate the reasons why that phenomenon allows so many different analytical descriptions, and how all these special pictures can be combined within an integrative paradigm.

Exploration of subject-determined or subject-oriented features in natural and cultural phenomena is the fundamental method of consciousness study. Natural things will definitely have other aspects, originally irrelevant to subjectivity; it is important to properly distinguish natural behavior from subjectively mediated. All kinds of things can be thus analyzed (and made objects rather than things in themselves); the subject's self-reflection requires contemplating things as they are used by the subject.

Since scientific research is also an activity, one can abstract its subjective component as well, deriving the general properties of consciousness from the very structure of science and the history of its development. This leads to the so called "second order" sciences, studying the Subject's involvement in the objective processes described by the corresponding "first order" science. Such studies are related to methodology, which provides the necessary background for "second order" science; however, "second order" research

---

[95] The relations between objectified and subjective knowledge can even take the form of contradiction; this contradiction is resolved in a the growth of the hierarchy of knowledge in general.

remains mainly scientific, while methodology (even in the form of metascience) is rather a part of philosophy.[96]

Iterating this self-reflection of science, we come to "higher order" sciences; the focus of research will shift still farther from the original range of problems, and closer to methodology of science.

The same "second-order" (reflexive) approach could be applied to any activity at all, not only scientific research. Thus, the arts can express their vision of artistic creativity; and philosophy will consider its subjective component, which results in the appearance of many individual philosophies.[97] Also people normally have some syncretic idea of how they think and feel—this is the common reflection of common consciousness. Moreover, subjective self-reflection is in no way bound to knowledge; there is a hierarchy of forms, from mere sharing consciousness with the others up to active shaping of its future forms. These all are the components of the integral picture of a rather specific object, the very essence of which is its mutability and flexibility, and different paradigms should be combined for an adequate description of phenomena that are diverse by their nature.

Speaking about scientific descriptions, one could ponder upon the contributions of the "natural science" and "humanitarian" components in our knowledge of consciousness. This traditional distinction must be made more specific in the hierarchical approach. The subject is a part of nature, and it has its natural properties to be studied by natural sciences. The subject is a part of the culture (as the "second", artificial nature)—and this is the domain of humanitarian science. Physical, chemical, biological *etc* research provides the important information about the level of complexity necessary for an object to support consciousness; complementarily, the description of the cultural diversity tells us what kind of functionality must be supported.

Psychology occupies a special place in the study of consciousness. While other sciences are centered on natural or cultural phenomena, providing information about consciousness as a by-product, the primary object of psychology apparently coincides with the subject; psychology claims to deal with subjectivity as such. For many people, psychologists are

---

[96] Second order science is a syncretic form of methodology. Like in any hierarchy, there are infinitely many intermediate levels between science and philosophy, and the attribution of a particular research to science or philosophy is relative, depending on the position of hierarchy (and hence the cultural context).

[97] For philosophy, learning its own history is of crucial importance; the variety of philosophical schools is the most direct way of objective observation of the fundamental philosophical categories.

to explain human soul, as opposed to the body, and human spirituality as opposed to material action.[98] In analogy to the triad $O \to S \to P$, psychology occupies the intermediate place between "physical" and "cultural" research. This does not mean that psychological methods are different the methods of other sciences; however, due to the special position of psychology in scientific study of consciousness, psychological methodology is very susceptible to the influences of other sciences, and its methods are mostly adapted from the objective and productive poles of the triad. The diversity of psychological methods resulted in the formation of numerous psychological schools; some of them are closer to the objective pole and natural sciences, while some others concentrate near the productive pole (culturology). As usual, psychological knowledge needs them all, integrated in the hierarchy of scientific psychology.

The special role of psychology in consciousness studies has yet another turn. Due to psychology's intermediate (mediating) position, every fact about consciousness must have psychological counterparts or components. Conversely, since the minds of researchers reflect the studied phenomena depending on their cultural environment, every piece of knowledge about consciousness will be psychologically acceptable, and observable in properly organized introspection.

Of course, psychology does not have the monopoly on studying the subject as their dominant interest. Among other sciences that can be considered as a formal description of the subject, one could mention the complementary areas of economy and history; the former is closer to the objective pole, the latter to the productive pole. However, the both sciences deal with the immediate material implementation of the collective subject, and hence their results are primary to any psychological study, which analyzes the interiorized forms of the collective phenomena discovered in history and economy.

Mathematics and formal logic could refer to the formal aspects of conscious activity in general and can therefore represent knowledge about subjectivity as such. The same general knowledge can be provided by a number of more special sciences like systems theory, cybernetics *etc*, as long as they apply to a wide range of activities.

"Natural" sciences can also be used in a reflective way, providing the schemes for interpreting consciousness-related phenomena on any level.

---

[98] Religion has been always trying to usurp the domain of spirituality; however, any religious belief (and dogmatism in general) assumes mental slavery, and hence lack of spirit.

Thus, the apparatus of Newtonian mechanics could be used as a scheme for the dynamics of motivation, or as a model of logical inference, which has nothing to do with the actual mechanical motion in the physical space[99]. The reverse, when the dynamics of consciousness applies to a physical system, can happen only in the context of evaluating the influence of the subject on physical processes. There is a principal difference between physics and psychology, or between biology and history; there objectively exist lower and higher levels, despite all the convertibility and reflection. The world *is* developing in reality, and the direction of this (virtually irreversible) development determines the distinction between the levels of any hierarchy.

Scientific and non-scientific study of consciousness is to discover the distinctive features of a conscious being, as compared to non-conscious things and creatures, as well as the manifestations of consciousness in the physical and biological world. It is only through the imprints produced by the subject on the world, as well as the subject's ability to recognize these imprints as distinct from mere physical or biological motion, that subjectivity as such can be comprehended.

## *Observing the subject*

Normally, experimental science organizes human activity so that the well-controlled conditions result in an outcome from a pre-defined range, as predicted by some theory.[100] The regularities observed in experiment give the necessary feedback for the theory to develop better models of the object area under consideration. It is not a trivial task, to prove that some experimental setup is adequate to explore the definite class of phenomena. Quite often, this requires yet another theory, which makes our observations indirect; most science is based on such indirect observations, which is a special case of tools/instruments mediated activity.

From the definition of the subject as universal mediation, it follows that

---

[99] It does not really matter which branch of mechanics is used. Each of them has its own area of applicability. The different aspects of conscious activity require different paradigms; some of them are better suited for the methods of analytical mechanics, while some others may well fit in the quantum picture. Both relativistic and non-relativistic schemes can find their application in physical psychology and other sciences. It is only in philosophy that the unity of all such partial schemes is established, and the integrity of the subject is thus restored.

[100] That theory does not need to be quantitative; all it must predict is that there are distinguishable outcomes, and they are somehow dependent on certain conditions.

direct observation of the subject is utterly impossible. All we can observe is the products of conscious activity, and by these products we judge about the subjective processes behind them. There is no direct introspection, and the only possible way of self-reflection is to analyze our own actions, both in the outer world and within the organic or inorganic body implementing our individual subjectivity.

Basically, the most general experimental method in consciousness studies is to initiate a hierarchical activity, when producing a definite effect on the ground level gets recorded on a higher (more reflexive) level. Due to essential reflexivity of the subject's self-exploration, it does not matter which activity will be chosen. However, there is a variety of special methods, depending on the specific relations between the ground activity and the way of reflection about it.

In the simplest case, the researcher can analyze his/her own actions in relevant situations and put forward a number of conjectures about the possible regularities. In this introspection, the same person assumes two different roles, being present at different levels of hierarchy at the same time. This possibility is akin to the very mechanism of consciousness formation, with the products of activity representing that activity as hierarchical objects. However, such observations are not restricted to an individual subject, or the subjects of the same level. For instance, an observer can consider the behavior of other individuals along with his/her own behavior; this provides essentially the same information. Observation and analysis of social processes belongs to the same group of techniques. The study of the historical forms of economic and social organization as the traces of subjectivity in the culture is also a sort of introspection.

Transition from mere observation to active experimenting implies using model activities, with the motives controlled by the experimenter. To make the internal mechanisms of one's behavior observable, the scientist will manipulate with the person's environment, inducing a specially designed hierarchy of motives, thus separating the activity of interest (test activity) from the probing activities used to exteriorize the results. Currently, this scheme is predominantly employed by psychology, but it is not the only possibility, and human history knows a number of social experiments,[101] though such experimenting is not always compatible with human ethics. However, the special position of psychology among the other sciences about

---

[101] In a way, such experiments can also be called psychological, though referring to the psychology of a collective rather than individual subject.

conscious behavior allows considering psychological experiment as a representative of experimenting on subjectivity in general, especially exploring the internal side of consciousness, the subjective in subjectivity. That is why further discussion will mostly be concerned with experimental methods in psychology.

The implementations of the motive control scheme can be most flexible. However, there is always a product serving as the link between the levels of activity, and communication between at least two subjects, not necessarily represented by different persons. When communication between the subjects of different levels is involved (*e.g.* a person and a group), or between the different levels of the same subject, the experiment should be organized to separate the roles of the subjects, so that the observer would remain on a higher level of activity during the experiment.

When the experimenter extracts some information from the controlled activity, this looks like a special interpretation, or a projection of the test activity onto some other activity, which is not always a simple procedure. The adequacy of such an interpretation must be checked comparing experimental schemes with different test activities.

Any psychological experiment employs some physical or physiological processes as the indicators of subjective events. No subjective event can occur without such "material" changes—however, care must be exercised in interpreting the results and determining what has actually been established in experiment, since there is no direct link between physical or physiological effects and the manifestations of consciousness. Physiological (*e.g.* neural) processes are not directly related to mental processes, and no observable behavior can be unambiguously interpreted as a manifestation of consciousness and subjectivity. The same behavioral patterns can be caused by quite different influences, and distinguishing a conscious act from non-conscious implies a careful analysis of the social and cultural context. There are no abstract "objective" criteria, which would reduce consciousness study to simple measurement; the productive aspect of any activity is to be taken into account as well. Thus, higher animals can behave in a very sensitive way, when living with people for many years; they may be even cleverer and more intelligent than some humans in certain aspects—however, this is not a reason for claiming them conscious and attributing human motivation and mentality to the animals. On the other hand, intelligent behavior in humans can sometimes be much less reasonable that apparently irrational and unwise conduct.

The results of a psychological experiment can be linked to the internal organization of psychological phenomena (including the psychological aspects of consciousness and the unconscious) employing such a fundamental feature of hierarchies as their convertibility. Any kind of outer behavior must correspond to an inner mechanism formed in the process of interiorization, folding the activity into actions and operations. Conversely, every internal act can be nothing but a folded activity, which could be drawn to the topmost level of hierarchy under carefully designed experimental conditions. For instance, a psychophysical experiment with a person determining the subjective pitch of a tone may explicate the internal representation of a simple tone as a standard statistical distribution—if the same distribution is discovered in a different sensory modality, it could be quite logically deduced that this activity would fold into operations and actions similar to those already known in pitch perception, and the same formalism could be used to describe different classes of higher-level phenomena.[102]

Communication between the experimenter and the examinee modifies their behavior and influences their motivation structures. Consequently, the results are necessarily biased. Unlike in many other sciences, this interference cannot be made negligible, and the very relevance of the results to subjectivity depends on the examinee's acting as a subject, and not a mere object. Thus, projective tests and psychoanalysis can provide most valuable psychological information about the subject, but their application essentially depends on the personality of the analyst, and the more neutral is experimenter's attitude to the examinee, the more scarce and trivial the results are.

Studying the hierarchical organization of the subject requires special experimental techniques. Hierarchies grow in the process of development; this means that the model activity used to reveal certain aspects of consciousness has to be complex enough to allow personal development in the course of a single experiment. The general principles of hierarchical approach indicate that the possible solutions would involve several interacting activities, and their interference organized in a controllable way; the analysis of the products of these activities allows reconstruction of a number of hierarchical structures, representing the hierarchy of the subject, in its specific manifestations. However, such interpretation also depends on

---

[102] P .B. Ivanov "A Hierarchical Theory of Aesthetic Perception: Scales in the Visual Arts" *Leonardo Music Journal*, **5**, 49–55 (1995)

the motives and personality of the experimenter; there is no practical use of depersonalized analysis. That is, the results of the experiment will not describe individual development, if there is too little involvement of the experimenter, and the experimenter does not develop together with the examinees. In this case, the experimental setup is reduced to mere observation of the behavior of some (collective) subject, albeit in the artificial conditions. Despite the possible usefulness of such syncretic experimenting, the observer must be interested in the results of the test activities to explicate the deeper mechanisms of subjective development. There are many ways to achieve this; for instance, a fundamental theory can shape the preferences of the experimenter and provide the necessary social background for consciousness study.

### *Extrasensory reflection?*

Since the level of subjectivity is qualitatively different from those of inanimate existence and life, one might expect that the interaction of a conscious being with the world may involve, along with the lower-level laws, some specific influences, so that some changes in the world might be "marked" by the signs of conscious intervention, as distinguished from the natural "background". Such subtle effects are often attributed to direct influence of consciousness on the physical world, commonly known as extrasensory (or paranormal) phenomena or experiences. However, while there is no clear understanding of what consciousness really is, it is difficult to say whether a particular phenomenon should be attributed to conscious effort or some hidden natural mechanisms. Whether the possible effects of that kind can be used in scientific study of consciousness is still an open question.[103]

In the hierarchical approach, the apparently direct influence of consciousness onto the world is obviously related to highly folded activity, forming virtual links that look as a material effect of a mental act. Such phenomena cannot be entirely explained with the natural laws, but they remain mere manifestations of some less common aspects of quite common processes.

---

[103] Unfortunately, this area has been significantly compromised by exaggerated fantasies, dishonest speculations and faked sensations; today, a serious scientist is reluctant to consider the problems like that.

One possible explanation for unusual correlations seemingly observed in experiments on extrasensory perception comes from the notion of a collective effect, well known in physics and other natural science, but much less frequently employed in biosciences and humanities. Two persons can act in synchrony simply because they are involved in the same activity, and they do not need to physically communicate, to maintain this correlation for a long time, since the very structure of activity (which is a part of the culture) implies quite definite role behavior. Similar effects could be observed in non-conscious systems involved in human activities, as long as they behave differently from what would be expected in a natural environment, without the subject's interference. Still, the possibility of hidden correlations of a conscious individual with a physical system (like electroencephalograph) must certainly be treated with caution, to avoid inappropriate attribution to consciousness in the situations, where merely physical and physiological factors would be more appropriate.

## *Theories of consciousness*

Like any other science, sciences studying consciousness can use all the variety of theoretical models, including descriptive, empirical and semi-empirical, dynamical, statistical *etc*. However, since formal models necessarily abstract the object from its description, their applicability to studying the essentially reflexive phenomena related to consciousness is much more limited than in any other domain. That is, the formally obtained results cannot be trusted on the basis of mere logic of the theory, and they will remain mere hypothesis until there is a clear indication of the cultural situation requiring that very type of behavior. Interpretation of theoretical constructs becomes an important part of theorizing.

On the other hand, the reflexivity of the subject's self-comprehension implies that practically any theoretical construct can describe some aspects of conscious activity. If there is nothing in the culture that would correspond to a formally obtained result, people can design it, creating a new sphere of material production or social relations following theoretical prescriptions. There are no "good" or "bad" theories; there is only appropriate or inappropriate application of a theory.

In particular, theories describing non-conscious existence or life can be formally transferred to consciousness studies. The usage of such lower-level models is possible because subjectivity as universal mediation encapsulates all the other kinds of mediation, and the laws of physical motion must be represented in it as well as the laws of life. The converse is not true, and the lower levels do not imply any higher-level interference: elementary particles and atoms can move without any relation to life or consciousness, and, in general, live organisms exist prior to consciousness and can well do without it. Of course, the subject can influence physical or biological processes, involving them into conscious activity, so that the lower-level phenomena would acquire specific quality, distinguishing them from similar natural processes. However, this modification of natural motion can only be mediated by appropriate tools converting conscious actions into objective effect of the corresponding level, and consciousness always acts on nature according to natural laws. This means that one does not need consciousness to explain quantum phenomena, crystal growth, or DNA replication, though all these processes can be consciously controlled.

Subjectivity can be implemented in many ways, provided the universal schemes of conscious activity are reproduced in some material substrate. Accordingly, formal description of consciousness and activity is not limited to any particular approach, since the very universality of the subject indicates that any mental model reflects certain features of consciousness, while some other features will require a different approach. No model is better than another; every paradigm can be applied with equal success—and with similar restrictions.

Transfer of theoretical models from one domain to another implies reinterpretation of the basic notions. Thus, the apparatus of Newtonian mechanics could be used as a scheme for the dynamics of motivation, which has nothing to do with the actual mechanical motion in physical space. Similarly, using the elements of information theory or quantum mechanics in a theory of aesthetic perception does not reduce conscious behavior to mere information transfer or microscopic motion.

One or another theory may become preferential for description of a specific class of phenomena depending on the modes of their involvement in human activity. Thus, if there is a space-like parameter that can be measured continuously in time (or any other serial variable) classical mechanics is most likely to be applicable, correspondingly re-interpreting the quantities involved. For instance, a dominant motive of activity can be associated with a (multidimensional) space coordinate evolving in subjective time (measured by the number of reflection cycles) according to the "psychological forces," which reflect the external influences the person is subjected to. Similarly, a theory of the conscious control of body motion can be developed within analytical mechanics,[104] which is a natural way of describing complex many-body mechanical systems in physics. On the contrary, for a "scattering-like" experiment, when a person becomes subjected to some standard external stimuli with the person's reactions recorded and their dependencies on the parameters of the stimuli investigated, quantum mechanics would be more appropriate.[105] This is the most frequent (though not the only possible) type of psychological experiment. Similarly, experiments with relatively isolated personalities or the members of formal groups could well be described by non-relativistic physical models, while accounting for the person's communication on a larger scale, with the propagation of cultural phenomena involved, would require relativistic theory (for instance, studying socialization or psychological conflicts). For self-organization phenomena both in person and in groups, various models of chaotic dynamics are naturally applicable, while the influence of culture on the dynamics of personality is a possible domain of general relativity (including its quantum formulations).

Mathematics is a powerful tool for theory development. It allows constructing formal models of any complexity, fixing the structure of the object in an objective manner. However, since the idea of the structure refers to the static side of the whole, it becomes clear that mathematics is incompatible with any motion; this explains why mathematicians have been always trying to expel movement (and development) from the very language of mathematics, and even the "alternative" mathematical trends (like constructivism) speak of dynamics in a static way, imposed by the traditional forms of mathematical reasoning. The mathematical description of a process only refers to the *structure of the process*; accordingly, mathematical models of development mainly reflect its structural aspect. Theoretical explanation of motion and development is beyond the mathematical carcass of the theory.

The models of physical or other non-conscious systems can reveal the essential features of subjectivity due to the presence of fundamental regularities common to all the levels of reflection. There are physical and biological phenomena that could serve as prototypes of subjectivity and any theory of conscious behavior will incorporate them. These are *non-linearity* and *collective effects*.

---

[104] Г. В. Коренев *Введение в механику человека* (Москва: Наука, 1977)

[105] Ю. А. Ивлиев "Новые математические методы в психологии, их развитие и приложения" *Психологический журнал*, **9**, 103–113 (1988)

Non-linearity can enter theory in different ways. On the lowest level, *weak* non-linearity comes as a constraint on the dynamics of a linear system (initial conditions, boundary conditions, sources and sinks *etc*); on a higher level, we find *imposed* non-linearity, which is of high importance for theories describing observer-dependent phenomena: a spatially or temporally restricted zone of view makes linear dynamics appear non-linear, producing the observed effects that are not actually present in the dynamics of the system observed.

The next level is that of non-linear dynamics. The equations of motion may be explicitly non-linear (*strong* non-linearity), or they may contain varying parameters controlled by an external process (*induced* or *parametric* non-linearity).

Finally, one can consider non-linearity as an effect of global correlations on local motion. Its objective form is represented by *self-consistency* and *self-organization*. Thus, there are numerous non-linear effects in both quantum and classical thermodynamics and kinetics, which lead to the variety of phase transition phenomena, chaos and catastrophes. The observer-mediated forms include positive and negative feed-back, averaging *etc*.

In a complex enough non-linear system, there are modes of motion, when most distant parts of the system move in synchrony, though their synchronization cannot be explained by direct interaction of the parts. For instance, a standing wave between two rigid boundaries is characterized by correlated oscillation phases on the opposite boundaries, though the time of the propagation of a perturbation from one boundary to another may be much greater than the period of vibration; this system is only weakly non-linear, with non-linearity being introduced through the boundary conditions. In richer systems, many partial waves can interfere in a complex manner, producing most intricate patterns of motion. One could mention solitons in hydrodynamics, shock and stress waves in solid-state physics, wave packets, phonons, laser modes, autoionizing states *etc* in quantum physics. Numerous examples could also be drawn from chemistry and biology. Very complex timbres of musical instruments are the most obvious example from an area other than science.

Social systems can behave in a relatively simple way in certain respects—and they can be described by linear or weakly non-linear models in that case. However, the very existence of the society as an integral formation is based on reflection, which implies non-linearity on all levels. Various social processes may interfere in such a way as to produce relatively stable formations of different scope. These collective effects are the individualizations of the abstract (universal) subjectivity; they appear to exist on themselves, relatively independent of the society that produced them—up to the degree of claiming to be entirely free of it. Such a view naturally follows from the fact that the universal mediation effectuated by large social groups is less apparent, and it can be too global for the individuals to notice it. However, in the dramatic periods of human history, the activity of masses became quite noticeable, though afterwards it might become subjectively ascribed to the will of an eminent person, a genius. This is one more feature of essentially non-linear systems: every global effect must be reflected in the dynamics of quasi-stable local formations.[106] Many psychological and social phenomena can be easily interpreted within the collective-effect model of subjectivity, and one could compare consciousness to correlations in non-linear medium resulting in a resonance shape. The most important of them is the unity of apparent individuality and inherent sociality in every person, which can hardly be explained any other way.

To summarize, collective effects in social systems produce quasi-stable hierarchical structures that can be called individual subjects; by definition, the structure and behavior of such individuals depends on the social conditions they live in. The topmost element of the hierarchical structure (like the crest of the wave), defining the individual subject is most often, though not necessarily, centered on a representative of the biological species *homo sapiens*. The projection of this hierarchy onto its elements is called consciousness.

## Cultural dependence of science

In the hierarchical philosophy of consciousness and subjectivity, science finds its place in the hierarchy of the forms of reflection as a level of analytical creativity intermediate between art and philosophy. In this context, we speak about creativity as an activity of reflection separate from the reflected activity, and the results of reflection (its product) are different from the product of the reflected activity. All forms of creativity are analytical in

---

[106] This reminds the particle-wave duality in quantum physics, and especially quantum field theory, where every field is quantized to behave as an ensemble of particles, and every particle is just a manifestation of a field.

this sense. However, within creativity, one can distinguish syncretic, analytical and synthetic forms, depending on the way the object and subject of creative activity are interrelated. In the arts, the subject and the object penetrate each other, and the artistic product presents a subjective vision of the object; in science, we try to present everything in an objective manner, stressing the difference of a thing as it is from the forms of its reflection.

As analytical creativity, science is represented in culture by special institutions, in both formal and non-formalized (traditional) forms. This institutionalized science obviously depends on the current level of cultural development in general, and the dominating ideology in particular. However, even in the folded form of individual scientific picture of the world, science contains all the cultural dependencies of the institutionalized existence. Both the content and the form of science are thus socially controlled.

Being a kind of activity (external or internal), science obeys the same cultural restrictions as any other activity, and develops with the development of the society.[107] As institutionalized reflection, science belongs to the culture and it cannot give more than it is possible within the current level of economic and social development. Though every scientific result bears some universality (which philosophers tend to mistakenly call *absolute truth*), it can only be *implicit* in the body of relative and culture-dependent knowledge. As with an individual subject, who is only representing a social formation, the universal content of science is not in the scientific formulation of the results, but in the integrity of the cultural context. An individual scientific result only *represents* a piece of truth, which is hierarchical, including relatively stable and situational components. Still, this does not deny any objectivity at all, since the development of culture is an objective process too, and the forms of activity can be scientifically studied, as well as their relation to the forms of thought.

However, no formal organization of research can guarantee the truth of results obtained and their interpretation. It is only in practical activity purposefully rearranging the world that the adequacy of a particular scientific model is established. The development of science always follows practical needs, being shaped by their reflection in (social) experience. Since the possible forms of practical experience are related to the objective aspect of the world, formal manipulations may often lead to sound hypotheses in

---

[107] The fact that any knowledge (and any conscious experience in general) is primarily an activity, and hence it reflects the current state and demands of the society, was overlooked by both vulgar materialism and idealism seeking for absolute truths for all times.

science, but the success of such "science within science" should not be overestimated.

The methods and principles of any science originate from the overall cultural situation and the current stage of economic and social development. The more so for sciences studying various aspects of consciousness, since people's reflection about themselves can essentially influence their social status and behavior. For instance, a theory postulating that conscious activity is nothing but biological functioning (or even a kind of "computing") supports the spirit of enterprise in those in a favorable social position, while killing any desire to do anything in those whose social position is unfavorable; the latter become less apt change the society towards more uniformity. The opposite view, assuming an entirely mystical origin of consciousness, leads to apparently the same results, though on a different psychological grounds: those deprived and oppressed can feel themselves in a preferable position requiring no active behavior, while social dominance is considered a kind of guilt. As a result, scientists themselves get influenced by ideological prejudice and fail to see the obvious solutions if such solutions violate the socially imposed paradigm. Quite naturally, such theoretical faults influence the practical applications of science. Thus class-oriented psychology leads to class-oriented therapy, and the patient's problems are misinterpreted, with negative (if not aggravating) treatment results. That is why remission rates for many mental disorders are still rather high, since their treatment does not remove the cause of the disease, merely disguising the disorder with superficially correct behavior, which, however, won't survive the slightest communicative instability.

Science belongs to its time, and scientists depend on the ideology of the class and prejudice of the social group. However, in many cases, such a limited science will be quite successful, since the very criteria of success depend on the social and cultural conditions. The selected paradigms originate from practical needs, and hence they will be efficient for some time, until the new social perspectives ripen up. When novel modes of production demand a different hierarchy of social values, the necessity of changing the dominant paradigms is felt as a crisis of science. Philosophy becomes much more important in such periods of change, and scientists become susceptible to what they would otherwise disdain as idle talk.

Each major period of social development (a cultural formation) has its own forms of analytical reflection, similar to the organization of production in general. One of the principal achievements of capitalism is formally liberating the individual, detaching people from their immediate social environment and from each other.[108] This is an aspect of the system of universal estrangement, opposing people to the products of their activity and the means of production, and hence to the society. That is why, for a bourgeois psychologist, a person does not differ from his or her biological body, and the boundary between the conscious being and the animal becomes entirely erased.[109] The boundary of *the Self* in a group is then understood as purely biological, bodily separation; this boundary is assumed to be fixed once and for ever, admitting no personal growth.

Similarly, the idea of a group subject does not fit in capitalist economy, and any control of an individual's activity by the group is treated as mere suppression of one's abstract freedom (unrestricted individualism). In psychotherapy this position results in poor adaptability of self-regulation techniques learned in laboratory conditions to real life; the only durable effect can arise from various manipulation technologies, which can hardly be considered as truly human behavior. Yes, manipulation can be successful and profitable, it can be very efficient—but it can never be psychologically comforting and resolving internal conflicts.

For a bourgeois psychologist, even the methods of group psychotherapy are centered on the individual, the group only providing the means of intentionally structuring the person's environment. A stable therapeutic effect can only be achieved if it is the group that is treated, but not the individuals within the group.

From the hierarchical viewpoint, an individual's psyche is a manifestation of the spirit in general, and the boundaries of *the self* are much wider than the biological body, belonging to the cultural environment; the extent of one's personality is determined by the hierarchy of the person's relation with the other people. The pleasure of the healthy and trained body and the well-functioning system of vegetative reactions is not enough for a conscious being—much more satisfaction comes from the feeling of one's social importance, the necessity for the humanity as a whole, no matter whether it will be accompanied by the open recognition, "success".

---

[108] Quite often this formal independence comes to pronounced antagonism; this is a necessary stage, to get aware of that abstract individualism and initiate the synthetic processes.

[109] In some cases this identifying conscious beings with animals is intentional as a part of brain-washing propaganda aimed at suppressing any ideological opposition on the subconscious level. Complementarily, there is a prejudice that consciousness is a mystical gift and humans are entirely different from animals, having nothing in common with them in the psychological domain.

Bourgeois psychology (and capitalist propaganda) presents momentary success as the final goal of any purposeful behavior, neglecting the higher (super-conscious) levels, or at most considering them as an inconvenience, obstacle, censorship, oppression. The economy of capitalism is primarily transaction based, and immediate profit will always overweigh more distant prospects. Similarly, bourgeois science prefers quick effect to solid construction, graphic achievements to prolonged support.[110] The hierarchical approach suggests that any significant changes in the structure of individual consciousness can only result from some as significant changes in one's cultural position; in particular, psychotherapy must help people to find the universal significance of their lives, which would not be as transient as quick success. In other words, one will see the *sense* of one's actions, and thus lay a solid foundation for a purposeful and creative life. The solution of individual problems is in including the individual in a global social and cultural process, thus opening the awareness of the actual place of the individual in the world, eliminating fear and supporting clear vision of the current possibilities. Such self-comprehension cannot be given by merely mastering a few manipulation techniques; quite often, psychotherapy is not enough to achieve this.

Under capitalism, the all-penetrating universality of estrangement leads to the reduction of the personality to an abstract social function.[111] This is the objective consequence of the economy based on the division of labor, which is a necessary stage in economic development corresponding to the level of productivity, when individuals cannot live without assimilating the others' products, but they are not yet unique enough to make the culture essentially dependent on them. In such a society, one cannot expect much personal integrity or well organized inter-personal relations; therefore, people will tend to doubt whether the others take them for what they pretend to be, and whether they have managed to be for the others as they intended. People feel that the others perceive them in an abstract way, as mere carriers of some economic or social functions. But the core of subjectivity is in the universality, sociality and even cosmic significance of the simple everyday activities—and the feeling of self-respect is entirely based on that global importance of each individual. In a well organized society, the very social

---

[110] Under capitalism, a scientist has to spend much time and effort on silly advertizing, since those who can fund scientific research do not bother about fundamental ideas, they only appreciate the promise of profit.

[111] Mere pronoun instead of personality!

organization helps people to discover their universal nature; under capitalism, one needs a professional psychotherapist to compensate for personality distortions.

## Scheme Transfer: Physical Psychology

Any natural science can be used as a model (paradigm) in consciousness studies. As a rule, the formal scheme of one science can be transferred to another with minor adaptations; the most difficult part is to properly reinterpret the formalism, avoiding all kinds of reductionism. In this section, I will discuss the relations between physics and psychology as an example of such scheme transfer.

This choice is explained by the common view of physics and psychology as the most typical representatives of "natural" and "humanitarian" science respectively. For brevity, all kinds of sciences about non-animated nature together with their metaphysical generalizations are often called "physics". Similarly, all sciences related to human behavior and consciousness are commonly meant under "psychology", ranging from neurophysiology to philosophical phenomenology. Instead, one can more specifically compare any particular branch of psychology with a peculiar area of physics; this will require a projection of the general discussion.

Currently, the interrelations between physics and psychology are vividly discussed in the literature, which indicates both the actuality of the topic as well as continuing lack of understanding. Most generally, there are three groups of questions:

1. What can psychology give to physics?
2. What can physics give to psychology?
3. Is there any way to combine these sciences?

In the modern literature, the first group of questions is primarily represented by introducing observer in quantum mechanics; the attempts to reinterpret statistical physics on the basis of subjective information constitute yet another major activity in this area. However, as an alternative to such artificial problems, there are other possible applications of psychological concepts in physics. Since things in a culturally modified environment behave differently from the isolated things, some future branches of physics could investigate the psychical aspects of subjectively mediated processes;

on the other hand, since life and consciousness take their origin in the inanimate world, the description of reflexive phenomena in physics (such as nonlinearity, collective effects *etc*) can borrow ideas and paradigms from psychology as well.

Here, I do not discuss the applications of the hierarchical approach to physics, and hence the principal focus of this section will be on the second group of problems. From the very beginning, I accept that physics and psychology are different sciences, despite the possible mutual influences. They describe different sides of reality and any admixture of the subject in physics is illegal within scientific methodology, as well as any reduction of psychology to physics. For instance, the debate on which physics is more appropriate for describing consciousness (quantum or classical mechanics, relativistic or nonrelativistic theory, *etc*) is meaningless; no kind of physics describes psychological phenomena, while all kinds of physics can equally be made paradigms for psychological study.

Nature is a hierarchy of objects, and each level of this hierarchy should be studied with methods appropriate at this level, so that the hierarchy of sciences reflects the natural hierarchy of the world. Thus, physics studies *physical objects* that are different from *psychological objects*; still, the both kinds of objects exist in nature independently of whether somebody is studying them or not. The development of any science is the process of simultaneous formation of its subject and its methods.

The hierarchy of nature is not rigid; it manifests itself as different hierarchical structures, and the levels distinguished in one structure may be fused together in another, and *vice versa*. Every two levels of the hierarchy imply an intermediate level, combining the features of the both. In science, it means that for every two sciences one may construct another science, lying "between" these two. In particular, one may seek for some combination of physics and psychology, which, evidently, is not unique since there can exist sciences intermediate for this combination and the "pure" physics or psychology.

The levels of hierarchy are *qualitatively* different, and one level cannot be reduced to another, or deduced from the other levels. In particular, psychological phenomena cannot be reduced to physiology, chemistry or physics, or deduced from them. Human psychology is drastically dependent on social factors, and consciousness must be considered as a collective effect arising from thousands of communication acts between many people rather than from some neural or physical processes in one's brain. However, consciousness would be impossible without appropriate material premises; one of which is the admirable versatility of the human brain.

## *Observer in physics*

The beginning of the twentieth century was marked by the appearance of two famous physical theories which seem to picture the Universe quite differently compared to the earlier conceptions. Up to now, relativity and quantum mechanics remain strange enough to excite popular admiration. However, quantum physics seems to be harder for intuition than relativism. People have gradually grown accustomed to the contracting measures and the curved space-time; but they cannot generally comprehend the transition from quantum waves to the solid definiteness of the perceptible world. This transition is often attributed to some conscious intervention, and the observer is claimed to be an indispensable part of quantum science.

But are quantum and classical theories as different as they appear? The difference will be less dramatic, considering numerous modern reformulations of both classical and quantum mechanics, which demonstrate rather smooth mutual transitions from one to another. So, is there any real need in the observer?

From the practical side, the task of a physicist is to predict the results of certain manipulations with material bodies using some pre-defined procedures called *physical methods*. The registration of some result follows a formal scheme which is called *measurement*. When an experimenter reports his results to the physical community, the main effort is spent to making the experimental procedure as close to a common standard as possible, to reduce the influence of any side factors, including experimenter's mind and personality. Thus the very idea of physical measurement assumes the extinction of the observer; this equally holds for classical and quantum measurements. The physical theory refines the schemes of measurement abstracting from the last traces of individuality; the whole of physics becomes then fully observerless.

The same idea can be put another way. Physical science deals with some *formal model* of observer, rather than a real human being, and it is this model that shapes the physical theory. Such a formal observer is just a representation of some standard procedure, and the observer's activity is reduced to the implementation of a sequence of operations, which could be much better performed by some automatic device. In this sense, the observer is present in any part of physics, and not only in quantum mechanics. It is only the rules of observation that change from one physical science to another depending on their specific methods.

Apparently the most objective of the physical theories, classical mechanics, is also dependent on the notion of a formal observer. There are many formulations of classical mechanics, and each formulation assumes its own way of postulating the formal behavior of the observer; still, one might say that all these observers are physically equivalent since they lead to the same dynamics. For example, the traditional Newtonian mechanics introduces the formal observer trough the idea of a *reference frame*. To observe the classical behavior of a physical system, the observer should be present in any point of the three-dimensional space in the same moment of time, to become aware of any event immediately. Such an observer is effectively infinite and coincides with the whole of the Universe. This picture can only be adequate if the movements to consider are much slower than the movements involved in the process of measurement (the adiabatic limit).

The relativistic generalization of classical mechanics is obtained when the movements to describe are as fast as, or even faster than the processes implied by the measurement scheme. To keep the notion of the reference frame, physicists have to associate it with the own movement of the observer, thus mixing space and time in the four-dimensional space-time. In other words, the reference frame is not a static prerequisite, but rather the process of establishing the connection between different spatial points. Relativistic observer is essentially local and cannot be aware of events occurring in very far spatial points.

Quantum mechanics generalizes the classical ideas in another direction. While relativism speaks of a very small observer, the observer of quantum theory is extremely big, even much bigger than the classical (infinite) observer. Each point of its space (a reference frame) becomes a whole three-dimensional space, and each point of this *internal* space is supposed to be somehow structured too, when it comes to accounting for spin and other intrinsic *symmetries*. For example, the infinity of atomic physics is practically about several microns, or even fractions of a micron, which can be considered a point in many macroscopic movements. A mathematical theory idealizes this scale difference, taking the *practical* infinity for true infinity; this is the source of formal contradictions arising when one tries to comprehend the transition from quantum processes to macroscopic measurements, from one level of hierarchy to another.

Normally, physicists clearly understand the limited applicability of theoretical abstractions and easily switch from one theory to another depending on the circumstances. Thus, the region between atomic and macroscopic lengths is better described by quasi-classical approach, and the same nucleus may be considered either as a solid body, or as a Fermi gas, or as a quantum liquid. Sometimes, however, theoreticians would forget about the real meaning of their terms and raise a controversy about nothing.[112] Unfortunately, many popular relations of physical theories lack indications to the limits of their applicability, so that the readers are easily deluded by some peremptory claims.

In the same way, the abstraction of an observer might be discovered behind any other branch of physics, like thermodynamics, electromagnetism, hydrodynamics, and so on. Similarly, there are natural transitions between physical sciences. For example, adiabatic processes in thermodynamics manifest quite classical behavior, and the phenomenological parameters like temperature, volume, or pressure, can be used as generalized coordinates. Lack of adiabaticity requires a quasi-quantum or quasi-relativistic approach. In all cases, the abstract observer of a physical theory does not imply any direct interfering of a human being with a physical system; all one needs to do is to *prepare* the physical system to behave in a definite manner, while the physical processes themselves are independent of the observer. Moreover, the conditions for certain types of physical motion can occur in the world without any conscious activity and prior to any subjectivity. Rather, the possibility of practically establishing some physical laws is based on the inherent possibilities of the culture within a physical world.

### *Physical methods in psychology*

Given that physics provides a variety of abstractions to describe idealized actions of a human observer, one might ask whether such formal descriptions could be useful in studying human behavior in general, rather than mere physical experimenting. For instance, quantum or classical mechanics might reflect some features common to a wide range of human activities, or even some universal traits. After all, science begins where unique events become generalized and thus made the abstract schemes applicable to many particular

---

[112] In such cases, they do not act as physicists, rather exercising some home-made (or ideologically suggested) philosophy.

cases. Psychology, if it wants to be a science, has to develop its own abstractions, and one cannot demand it to give a comprehensive explanation of any detail of a single behavioral act. On the contrary, psychological analysis is aimed at classifying individual acts, bringing then under some predefined categories, which are familiar enough to enable people's control over their own behavior, just like people control physical processes.

In complement to inventing theories from scratch, psychologists can take a ready-made formal scheme from physics (or another science) and apply it to psychological phenomena. Of course, no physics can *explain* psychological phenomena and consciousness—this is the task of psychology proper. Likewise, psychology cannot be derived from any chemical or biological laws, from the physiology of the brain or computer analogies. All what is legal to ask is how these biological, chemical or physical processes are involved in a conscious action, as soon as one knows that they are indeed involved.

There are different ways of approaching psychology from the physical side. One approach is to treat a human being as a physical object, albeit very unusual one. Then we can physically act on that object and observe its physical reactions, trying to catch the apparent regularities in some formulas. For instance, exposing a person to some flashing lights, various sounds, electric shocks, sequences of words or even congruous texts, music or movies, may result in person's reactions, like pressing a button, saying something, going to a nearby supermarket and buying a new hat, and so on. Physical measurement does not worry about the specific personal sense of these reactions; all that is relevant is distinguishing a number of *physically different* outcomes which can be somehow labeled with a numerical parameter. Such procedures can be highly formalized, and they differ from physical experiments proper only in their object. This is in the least degree a psychological study, and it can as well be considered as a kind of physical research, namely *psychological physics*, or *psychophysics*. The most popular psychophysical methods include various timing procedures, the measurements of transmission characteristics (for example, the dependence of evaluated sound pitch on the sound frequency), and numerous threshold measurements (quite analogous, say, to ionization potential measurements in atomic physics).

Here, the key point is the usage of *physical* criteria for distinguishing different reactions. Thus, if one is interested in physiological consequences of some manipulations with a human being, this is psychophysiological rather than psychophysical study. Likewise, one might consider the influence of stress onto speech production—this is a psycholinguistic study. Psychological research proper would be interested in the specifically psychological phenomena, such as changes in the motivation structure or the level of self-respect. It does not matter how formal the means of this study are, as long as the focus is on the psychological side of the problem. Psychological notions are not a bit less abstract than the most abstruse constructions in theoretical physics. "The will" may seem more familiar than "autoionization", but this is mostly due to more evident manifestation of the will in our everyday life, while autoionization (like any other microscopic event) is not so easily observed, despite its being much more common in the world.

There is an important distinction between the higher and lower levels of hierarchy. Any psychological event can always be considered from the physical side as a sequence of physical events, while there are physical events that do not assume any psychological content, and the same physical events can accompany many psychological events. However, since human knowledge about the physical world is governed by people's practical needs, science only deals with events related to human activities, and, in principle, one could reveal some relation to psychology in any *known* physical event.

Psychophysics is not the only way to combine physics and psychology. Since any movement in the mind is implemented in a sequence of physical events, mental phenomena cannot violate physical laws; this, in particular, makes it possible to predict some gross thought regularities common to all kinds of conscious being, including hypothetical creatures from imaginary worlds with different values of some fundamental physical constants (and hence a quite unusual physics). The structure of the physical world influence mental processes,[113] and this is yet another possible direction of boundary studies between physics and psychology.

People construct their models of reality according to their ways of acting in that reality. For instance, the human organism (like the majority of higher animals on the Earth) reacts on the second derivative of any physical process. For instance, we do not feel steady motion, or constant values or slow steady change in temperature, humidity *etc*—though we immediately know about any non-homogeneity in such changes. One of the results is that physics puts second-order equations in a preferential position. Similarly, the forms of

---

[113] F. J. Dyson "Time without end: Physics and biology in an open universe" *Reviews of Modern Physics*, **51**, 447–460 (1979)

conscious activity can determine the form of physical theories.

One more possible combination is a "compound" theory, where the influence of mind on physical movements is introduced explicitly as some phenomenological *constraint*, and, conversely, physical laws are regarded as the constraints for the possible changes of mind. Unfortunately, this approach did not attract many scientists, because the construction of such a theory requires huge technical work, which did not seem feasible until recently, without advanced means of computer modeling.

Now, let us imagine a typically psychological research leading to results that strongly resemble the behavior of some physical system. Nothing prevents the researcher from following this analogy as far as possible and applying the standard formalism of the corresponding physical theory to the regularities observed at the psychological level. This seems even more admissible since physics itself has been extensively practicing such formal borrowing of paradigms from other domains since pre-historic times. One can hardly find a physical theory that had not ever been influenced by some other science, either physical or not.

Originally, transfer of physical and mathematical theories to psychology is rather superfluous, with physics taken as a source of useful metaphors. However, there is a whole range of theories intermediate between such metaphorical usage and predictive theories based on the equations of dynamics.

Modern physics is rather broad-flung; it includes models far from the traditional dynamical approach. Fractals and neural networks have become very popular in the physics of condensed media and surfaces; also, there are theories examining computational or informational properties of physical systems. Quite probably, such boundary disciplines will lead to a new revolution in physics—still, they can be applied to psychological problems along with any other models to obtain an explanation of existing mental structures. One such model, combining quantum-mechanics and information theory with the ideas of the hierarchical approach has lead to a new theory of aesthetic perception, opening broad perspectives for both theoretical aesthetics and practical applications in the arts.[114] The model explains the discrete nature of musical pitch perception, so that the properties of all existing and new (theoretically possible) musical scales could be described with a few *a priori* computable values. Similar scaling has been discovered

---

[114] L. V. Avdeev and P. B. Ivanov "A Mathematical Model of Scale Perception" *Journal of Moscow Physical Society*, 3, 331–353 (1993).

in visual perception.[115]

When some formal constructions are transferred from physics to a psychological problem, they do not change the psychological orientation of research in general. Since it is psychological phenomena that are to be described, the parameters and variables of the theory must be psychologically reinterpreted, losing any relation to their physical counterparts. Accordingly, the results formally obtained in this model are psychological, rather than physical. Such theory is a branch in psychology; this approach, in contrast to psychophysics, could be called *physical psychology*.

## *The foundations of physical psychology*

Physical psychology is to investigate the possibilities for adapting physical theories to psychological applications; this implies modification of the original theoretical scheme for better description of psychological phenomena.

A formal scheme transfer from one area to another can only be possible if there is objective similarity of methods used in the both areas. Luckily, some similarity is bound to exist for any two sciences, which follows from the fundamental principle of the unity of the world and the universality of subjective reflection (and scientific reflection in particular). More specifically, there are methodological parallels due to the universality of the structural, systemic and hierarchical aspects of any part of reality. Thus, the general scheme of the scientific experiment assumes registration of the object's response to a number of standard external influences. In this scheme, the object is considered as a *system* transforming some input into some output trough a sequence of internal states. In a very clear form, this approach manifests itself in the matrix formalism of quantum scattering theory, as well as in the stimulus-reaction scheme of the classical behaviorism. Another class of scientific methods comes from the structural approach; the main goal of a structural study is the explication of the internal integrity of the object, connecting its distinct parts into the whole, opposed to its environment. For example, one could recall the atomic paradigm in physics and Gestalt psychology. The combination of the two paradigms leads to considering the object's development, and the stages of this development

---

[115] P. B. Ivanov "A hierarchical theory of aesthetic perception: Scales in the visual arts" *Leonardo Music Journal*, **5**, 49–55 (1995).

become represented in the object as the levels of its intrinsic hierarchy.

Physical psychology does not aim at obtaining new psychological data, leaving that to psychology proper. The formal models of physical psychology must conform to existing psychological data and give reasonable results, as soon as the measurement of the parameters of the model is possible. Being generally subordinate to psychology, physical psychology is useful to clarify the meaning of the existing experimental procedures; it can also suggest new qualitative and quantitative characteristics. The schemes of physical psychology are to complement specifically psychological research; they cannot replace it, especially where no physical models are applicable.[116]

Also, physical psychology is not a branch of physics, since its domain differs from the scope of physical sciences. Physical psychology borrows formal models from physics, but it applies them to the systems of quite another type, in which the processes do not directly correspond to physical processes in a system of material bodies. One could say that physical psychology is the physics of the ideal, in contrast to the ordinary, "material" physics. But, since the ideal and the material are just the two sides of one reality, one should expect that some features of physical models in psychology would somehow manifest themselves in physical research proper, and there would be a way back, from psychology to physics.

For physical psychology, a person is not only a material body having a definite motion in the physical space-time. The main interest concentrates on the internal, subjective processes that cannot be characterized with reaction delays, sensory organ attenuation curves, spatial distribution of excitation in the brain and interactions of its parts, the mechanics of muscles *etc.* That is why the scope of physical psychology does not coincide with the scope of psychophysics, which describes exactly these external manifestations of psychic processes. In a way, this difference is similar to the difference of the physiology of higher neural processes and neurophysiology: the former studies the physiological mechanisms underlying psychological phenomena, while the latter describes these phenomena in terms of neural dynamics.

Since theoretical physics widely exploits mathematics of all kinds, it might be expected that the same mathematics would be applicable to ideal, psychological processes. This application, however, is different from that of mathematical psychology. The latter studies the possibility of correlating psychological entities with mathematical objects as such, the ways of

---

[116] This is like the methods of mathematical physics cannot replace physical research proper; however, they are quite efficient in relatively stable theories.

mathematization. Naturally, one or another mathematical representation is a necessary stage in the development of a physical model, but mathematics is only a background for physical theory, the principal concepts of which lie beyond any mathematics. In physical psychology mathematics is only introduced through a physical model, and does not require direct mathematical analysis of psychological data. For example, there are situations in physics, when the same model is described with different mathematics (like the Heisenberg and Schrödinger representations in non-relativistic quantum mechanics); when this model is transferred to psychology, all its mathematical forms are transferred with it, and may be used without any special reservations as soon as the analogs of the corresponding quantities are discovered. On the other side, the mathematical methods of psychology cannot always be related to any physical model.

To summarize, physical psychology has a definite scope different from that of psychophysics and mathematical psychology; it combines the elements of physics, mathematics and psychology without being reduced to either of them.

## *The scheme of Newtonian mechanics*

Classical mechanics plays a special role in physics. Being developed for a few centuries, it brought physicists huge experience of constructing mechanical models for thousands of special cases. There are numerous reformulations of classical mechanics, clarifying its relation to other physical sciences. This is why new physical theories are first applied to classical models, which is the best way to demonstrate the essence of a new approach.

Speaking of physical psychology as a new science, it would be natural to appeal to classical mechanics to get a general conception of how physical models might work in psychology. The commonly known Newtonian mechanics is the simplest mechanical theory; this is the first step studying physics in general. That is why I have chosen to illustrate the methodology of scheme transfer from physics to psychology with a psychological model built on the basis of Newtonian mechanics. Omitting any computational details, I will concentrate on the conceptual shift from physics to psychology, and on the ways of reinterpreting the formal results of classical mechanics.

To fix terminology, I will briefly describe the formalism of classical mechanics in the Newtonian formulation.

The basic objects of this theory are *material points*. Each material point is characterized by its *mass*, which is usually denoted with the letter $m$. For each material point, one can specify its *position* in some *configuration space*, which can be either the ordinary three-dimensional space or some abstract space of one or more dimensions. One can fix a *reference frame* in the configuration space, and the position of some material point is then defined with a set of numbers, which are called its *coordinates* in this frame. Usually, the position of a material point is said to be a vector, being characterized by both its absolute value (length) and its orientation in the configuration space. I will denote the position of a material point with the letter $\boldsymbol{x}$, where the boldface indicates that this is a vector, and the length of this vector will be denoted with the same letter $x$ (in the normal face). The movement of a material point is described by changing its position in the configuration space with time $t$; this movement is characterized with a definite *velocity*, described with a vector $\boldsymbol{v}$, which is formally obtained as the first derivative of $\boldsymbol{x}$ in time: $\boldsymbol{v} = d\boldsymbol{x}/dt$. The first derivative of $\boldsymbol{v}$ is called *acceleration* and denoted with the letter $\boldsymbol{a}$. Yet another important quantity is the material point's *momentum* $\boldsymbol{p}$ defined as the product of its mass and its velocity: $\boldsymbol{p} = m\boldsymbol{v}$. The principal law of Newtonian dynamics is then formulated as follows: the first derivative of $\boldsymbol{p}$ in time is a vector function $\boldsymbol{F}$ of time, material point's position, and its velocity:

$$d\boldsymbol{p}/dt = F(t, \boldsymbol{x}, \boldsymbol{v}).$$

The function $\boldsymbol{F}$ depends on the nature of the physical system concerned and is called *force*. The solution of this *equation of motion* gives the position of the material point at any moment of time, and all the other characteristics can be calculated knowing $x(t)$.

A mechanical system may consist of many material points. In this case, the force acting on any one of them depends also on the positions and velocities of other material points, and the law of system's dynamics (commonly known as the second law of Newton) becomes a system of equations, one for each material point in the system. However, such system can be treated as containing only one material point moving in the space of higher dimension. For example, two points in the ordinary space are characterized by six coordinates, which can be interpreted as a point in a six-dimensional space.

Usually, in Newtonian mechanics, masses do not depend on time, and Newton's second law can be rewritten as follows:

$$d\boldsymbol{p}/dt = d(m\boldsymbol{v})/dt = m \cdot d\boldsymbol{v}/dt = m\boldsymbol{a} = \boldsymbol{F},$$

that is, the force acting on a material point equals its acceleration multiplied by its mass.

When **F** does not explicitly depend on $t$, there exists such function $U(x)$ such that

$$E = mv^2/2 + U(x)$$

does not depend on time. The value $E$ is called the *total energy* of the system, and it is the sum of *kinetic energy* $mv^2/2$ and *potential energy* $U(x)$. Since potential energy depends only on the position in the configuration space, it can be considered as some potential field existing in this space as an independent entity. Systems, for which total energy remains constant, are called conservative; the fact of the constancy of the system's total energy is often referred to as the law of energy conservation. For conservative systems, $x(t)$ can also be retrieved from this equation

$$mv^2/2 + U(x) = const$$

One very important instance of a mechanical systems is described by the equation of motion

$$m\boldsymbol{a} = -m\omega^2 \cdot (\boldsymbol{x} - \boldsymbol{x_0}),$$

that is, the force is proportional to the displacement from some *equilibrium point* $x_0$, and directed always back to this point. Such equations describe a wide range of oscillations around the point $x_0$; the system obeying an equation of this type is called *harmonic oscillator*.

The simplest one-dimensional solution of harmonic oscillator equation is given by

$$x = x_0 + A \cdot cos(\omega t + \varphi)$$

that is, the material point repeatedly ($\omega/2\pi$ times per the unit of time) moves away from the equilibrium point, and then returns to it, moving on in the opposite direction. The constant $A$ is the amplitude, and the constant $\varphi$ the phase of oscillation. The potential energy in this system is given by the equation

$$U = m\omega^2(x - x_0)^2/2$$

which has the same form as the expression for kinetic energy, with the only replacement $v \to \omega(\boldsymbol{x} - \boldsymbol{x_0})$. The minimum of the potential energy is achieved at $x = x_0$, the equilibrium point.

There are more complex oscillatory solutions, when each component of vector $\boldsymbol{x}$ oscillates with its own amplitude and phase. For example, two-dimensional oscillations correspond to movement along an ellipse; circular movement around the equilibrium point is a special case of two-dimensional oscillation. For a point, moving along a circle around the equilibrium point $x_0$, the velocity and acceleration of the material point are constant in the absolute value, and it is only their orientation that changes. This means that not only the total energy is conserved, but both kinetic energy and potential energy are separately conserved.

## *Motivation and temperament*

According to the general psychological theory,[117] each human activity is governed by some motive and unfolded in a sequence of actions directed to specific goals. People are unaware of their motives, and it is their goals that are conscious. In the course of action, the motivation may change, so that one activity transforms into another. Sometimes, the former goals become motives, and a motive may become merely an intermediate goal.

Let us assume that, in certain situations, a motive can be represented by a point in some *motivation space*. In this model, the goals will belong to the same space, to enable transformation of goals into motives, and motives into goals. Any human activity is represented by a trajectory $x(t)$ in the motivation space, that is, by a sequence of points representing the current goals. The motive of this activity is naturally represented by some attracting center in the motivation space; the activity can thus be obtained as a solution of an equation of motion, similar to the second law of Newton in classical mechanics.

Within this analogy, the mass *m* of the material point corresponds to the internal inertia of mind, which is an important personal characteristic. The greater is the mass, the less readily the person yields to external influences (represented by some "mechanical" forces); also, such people tend to preserve any inner motion once it has been initiated. Velocity *v* naturally describes the rapidity and the direction of an action; this is an example of a characteristic that has no direct psychological analog, though it is quite compatible with the psychological conceptions. As for momentum $p = mv$, the corresponding psychological characteristic might be called the *persistence* of the activity, that is, its ability to preserve the same course in

---

[117] A. N. Leontiev *Activity, consciousness and personality* (Englewood Cliffs, NJ: Prentice Hall Inc., 1978)

spite of any deflecting forces. Quite naturally, highly inert individuals ($m$ is high) are more persistent in their activity; also, the higher the rate of activity is, the less noticeable is the influence of other activities on it.

In Newtonian dynamics, acceleration plays a special role. Any change in the state of motion assumes non-zero acceleration, and it is acceleration that is felt by a classical observer (frame of reference) as a mechanical event. For an observer moving without acceleration, the dynamics of any mechanical system is described with the same equations of motion, as for observer in peace. This means that all the reference frames moving without acceleration are mechanically equivalent; since motion without acceleration corresponds to the absence of forces, and the state of motion is preserved once it has been initiated, such motion is called inertial, and steadily moving reference frames are called inertial as well. One naturally comes to the hypothesis that the inner representation of the forces acting on the person corresponds to acceleration in the mechanical model of activity; it can be associated with people's subjective experiences.

Now, the overall picture of human activity is pictured as follows: a person's interaction with the world (including the person's body, and the brain) results in some distribution of forces in the motivation space of the person; these forces excite definite affects in the person, changing the state of motion (the rapidity of changing actions and goals, and the direction of this change).

The immediate consequence of this model is that the same force will excite weaker emotions in a person with higher inertia, since acceleration equals force divided by mass. This is the well known low emotionality of the people with phlegmatic temperament. Following this line, one could ask whether the other classical temperaments (sanguine, choleric, and melancholic) might have a mechanical explanation too.

The distinction of the four temperaments takes its origin in the Ancient Greek philosophy; it has been physiologically interpreted in Ivan Pavlov's theory of reflexes. The temperaments are characterized by three parameters: strength, mobility, and balance. Thus, the sanguine temperament corresponds to strong, mobile, and well-balanced nervous processes; the choleric temperament is poorly balanced, while the phlegmatic temperament lacks mobility; all the weak temperaments are called melancholic. The mechanical interpretation of these parameters of temperament can be given on the basis of the principal law of dynamics: force equals mass times acceleration, $F = ma$. Observe that the strength of temperament characterizes the degree of the person's sensitivity to external circumstances. In the mechanical language one will say that the environment acts with less force on a person with greater strength of temperament; that is, the absolute value of the force $F$ is inversely related to the temperament strength. The relation of mass $m$ to inertia (the inverse of mobility) has already been indicated. Quite naturally, balance is characterized by the value of acceleration: the completely balanced state of the system assumes zero acceleration (pure inertial motion).

With these assumptions, the sanguine temperament must be characterized with small $F$, which, for medium $m$, results in low accelerations $a$. Since the phlegmatic temperament is characterized with a significantly higher mass, even much greater forces cause rather low accelerations, and a phlegmatic person keeps balance in a wider range of situations. The opposite holds for the choleric temperament, which assumes low inertia and hence even a small force can break the balance in a choleric person. As for the melancholic temperament, it is mostly characterized with a rather great sensitivity to the processes in the environment, that is, with high values of $F$. The effect of high $F$ on the person's activity can be different, depending on the person's inertia, which corresponds to the empirical distinction of the three types of melancholic temperament. Inert individuals remain balanced in spite of their strong interactions with the world. Medium inner mass results in much more pronounced affective reactions. The weakest type of melancholic (a classical melancholic) is characterized with low inertia; this is an extremely vulnerable person, feeling the flood of emotions at any turn of the situation.

The mechanical treatment of temperaments differs from the traditional approach in that strength and balance are usually assumed to be individual constants, while their counterparts in the mechanical model, force and acceleration, are true dynamic variables, which may significantly change in the course of activity. One possible solution of this problem is to treat temperament as the averaged feature of activity, relating its parameters to the time-averaged values of force and acceleration. For many periodic and quasi-periodic modes of motion, the absolute value of force (acceleration) varies in a narrow range, only changing its orientation. In the simplest case of circular motion, the force and acceleration are constant, which complies with the traditional treatment of temperament.

This mechanical model of activity can be developed in detail, finding more analogies between physics and psychology. As an example of a more complex result, I would like to mention the possible application of this model to the description of neuroses. Normally, there are no inaccessible regions in the motivation space; for any given point there exists some trajectory (activity) containing this point. This is a consequence of the universality of subjective mediation. Nevertheless, a person's interaction with the world can sometimes result in a singular potential, breaking the simple topology of the motivation space. The well known Coulomb potential of a charged point is an example from physics; in this field potential energy assumes an infinite value at the position of the electrical charge. In such cases, activity can come very close to the point of singularity, but it will only move around it, never achieving this point. The existence of such forbidden areas in the motivation space corresponds to the clinical picture of neurosis. The mechanical model permits the description of different kinds of neuroses, depending on the singularity type. The immediate implication is that a neurosis cannot be overcome by the own activity of the person; the treatment of neuroses requires a change in the person's environment, which will remove the singularity from the motivation space.

This is an example of how simple mechanical conceptions can be introduced into psychology of activity to describe phenomena quite different from the original physics. Of course, the same physical model is also applicable to other areas of psychology. Thus, one could reinterpret the mechanical equations of motion to describe communication between people, interaction of social roles in a small group, and so on. Alternatively, other physical theories can be used to describe dynamics of motivation in the situations of uncertainty and socially induced choice. For instance, while the paradigm of classical mechanics characterizes an individual action by the momentary goal and persistence of activity, in the quantum model, the point in the classical configuration space will be replaced with some internal space, and the action will become a process in this internal space, resulting in a probabilistic outcome on the level of outer activity.

Though formal schemes transferred from physics can be useful for the description of conscious activity, the origin of consciousness cannot be discussed within physical psychology, where we can only indicate the place of consciousness within the adopted physical model. Thus, in the mechanical theory of activity, consciousness is associated with the level of action, and the person is not aware of the motive of activity. This is quite understandable, regarding the goal (the point in the motivation space) as a focus of awareness; the activity is then interpreted as the gradual shift of this focus from one goal to another. Since the points of minimum potential energy (representing the possible motives) do not, in general, lie on the trajectory of activity, the motives remain unconscious. This is especially evident in the case of circular motion, with the motive in the center of the circle, and the goals always equally distanced from the motive. To make the motive conscious, a special activity of *motivation* is needed, additionally deflecting the trajectory of activity towards its motive; such dissipative forces can also be treated within the mechanical model. In general, some activities will include motivational actions, and some will not, depending on whether the motive point lies on the trajectory of activity or not.

Better representation of consciousness requires more sophisticated physical models, involving controlled nonlinearity and collective motion. Collective phenomena, like solitons in liquids and solids, plasma pinches, autoionization states in atoms, and many others, appear due to the system's interaction with itself mediated by its environment. Every individual body in the Universe is bound to its environment in many ways, and consideration of an isolated system is possible only in abstraction. The more so for the human brain, which is an instance of device capable of performing universal mediation; that is, the ability to interact with the whole world is one of its distinctive features. Consciousness is essentially a collective effect arising from the variety of interpersonal communications. This collective nature is reflected in the organization of the human brain and the interplay of the neural processes accompanying human activity. However, on the phenomenal level, one does not need a detailed knowledge of the social organization and its projection on neural processes; a physical theory of nonlinear phenomena can be used to qualitatively describe conscious behavior and make trustable predictions.

# CONSCIOUSNESS IN PSYCHOLOGY

Psychology is a science about psychic phenomena. That is, first, it is not philosophy and hence cannot pretend to comprehend the whole of subjectivity; second, it is not about subjectivity in general, but rather about a specific class of phenomena that can exhibit traces of consciousness, or, at least, are closely related to it. Psychology does not deal only with humans; there is animal psychology, and some peculiarities in computer operation can already be considered from the psychological viewpoint. That is, the *psyche* (soul) is rather an attribute of the higher animal (with the Roman *anima* being the translation of Greek *psyche*), and consciousness enters psychology through the essential modification of the psychic life under social influence. An animal can demonstrate almost conscious behavior—and people often behave in an animal way.

On the other hand, any science at all represents a domain of conscious activity and conscious reflection, therefore describing some aspects of consciousness by the very history of its discoveries and faults. These abstract facets of subjectivity are synthesized in philosophy to get an integral idea of the subject.

However, the idea of consciousness most commonly comes to people through the feeling of their selves, their inner motions and purposive behavior. That is why human psychology must be considered in any philosophy of consciousness, being a kind of natural test case and validity check. Of course, since philosophy is not science, no philosophy of consciousness can replace specifically psychological research. Philosophers feed on the facts of science, they can influence scientific observation, theory and experiment—but development of science follows its own ways. Scientific notions can grow to paradigms and become philosophical categories, thus leaving the realm of science. Conversely, in a special context, universal categories of philosophy can be interpreted in a restricted sense, thus shaping the form of an individual science or a particular scientific model.

From the general philosophical viewpoint, what could we want from psychology? In the hierarchical approach, every person (individual subject) is a hierarchy. An integral approach to the study of a human being must combine all the relevant manifestations of this hierarchy. Some of them will treat humans as mere physical bodies, chemical systems, or complex organisms. Understanding such aspects of one's individuality may give clues to the origin of certain peculiarities in one's manner of implementing consciousness, of being a subject. Psychology has its own niche in this hierarchy; namely, it deals with the integrity of inner structures and processes that maintains the integrity of outer behavior. That is, the basic unit of any psychological model is a public act, a unity of inner and outer aspects. Psychological regularities encapsulate any lower level laws; [118] there is no direct correspondence of any psychological characteristic to physiological or chemical reactions. A psychologist does not need to know the exact patterns of the brain's activity, or the details of muscular dynamics.[119] All one needs is a definite structure of the situation demanding inner decision to trigger outer activity.

This is an important point. Psychology starts where there is choice and communication. If somebody acts merely of strict necessity, there is no sense in asking why that way, and not another. Also, if the decision does not take the others into account, it must be treated as spontaneous, and there is no room for psychological analysis. Psychology could be roughly characterized as a science about the inner models of the outer world.

Today, psychologists normally work with individuals, trying to predict and regulate their behavior. Since only a few aspects of one's behavior are relevant in any situation, we do not need exhaustive knowledge of this particular person to make good predictions or influence one's actions in a specific respect. Human activity is hierarchical, and the higher level (gross)

---

[118] Human psychology could hence be characterized as a science about the ideal mediation of behavior.

[119] Of course, in very special cases, the observable behavior can result from a lower level feature (for instance, in stress, or malady). Also, there are boundary sciences between psychology and physiology (such as neuropsychology, or psychiatry) that require deep knowledge of the organic mechanisms underlying overall activity. Psychological knowledge there plays the role of a general context, a conceptual framework, allowing correct discrimination of the relevant physiological phenomena.

features of behavior depend on the lower level details in an integrative manner, with different low level states representing the same overall situation. Moreover, the different positions of hierarchy require different psychological techniques to cope with. This means that an adequate description of an individual requires a well-balanced combination of general and specific knowledge, and there is no single psychological model that would be useful in any case at all. Individuality does not mean that there are no universal laws. Conversely, individuality can only be observed as a specific manifestation of a general law. However, one always has to guess, which general laws are applicable to a particular individual. Psychological science provides a versatile collection of tools and instruments; their practical usage depends on the personal motives, and every case demands a unique combination of psychological scales and methods.

A practical psychologist has to be somewhat eclectic, inventing a unique evaluation technique for every individual, borrowing clues from anywhere, including sciences other than psychology, as well as non-science. However, for consistency, these diverse characteristics must comply with a general theoretical viewpoint, an integral idea of a conscious person. Speaking about somebody's psychology, we need to observe how all the variety of personal traits refers to a particular person as integrity.[120] This essentially depends on the ideological stand of the psychologist, conscious or not.[121]

However, the domain of psychology is not restricted to studying human individuals. Animal psychology and social psychology describes the active aspects of animal behavior. Social psychology includes such branches as group psychology, mass psychology, class psychology, national psychology *etc*. In the nearest future, computer psychology is certain to appear; quite probably, some extraterrestrial forms of *psyche* will be encountered some day. That is why it is important to have a wider view on the psychological aspects of subjectivity, relating its special categorial schemes to the general categorial hierarchies developed in the philosophy of consciousness.

Psychology has been historically developing as a variety of scientific schools and directions little compatible with each other. The numerous attempts to systematize psychological knowledge have not, so far, produced a consistent picture. Compared to the others, Hegel's hierarchy of the spirit

---

[120] This is what G. Allport called "systematic pluralism". Albeit based on the standard evaluation techniques, the final psychological scale takes shape during psychological study.

[121] Psychology cannot replace philosophy, or judge between philosophies. On the contrary, it necessarily makes use of some philosophy.

is, probably, the most fundamental and consistent, illustrating the basic method of developing any taxonomy from a unifying principle. The hierarchical philosophy based on the explicit idea of the unity of the world overcomes the idealistic bias of Hegel's philosophy considering it as one of the possible positions of the categorial hierarchy.[122]

Here, I start from the fundamental hierarchy of the forms of reflection

$$existence \to life \to activity,$$

which defines (conscious) activity as universal reflection, and primarily the synthesis of existence and life. Psychology refers to the level of life (with the soul animating the body, not necessarily organic); this means that considering the psychology of conscious activity requires a projection of activity onto life, and hence partial reproduction of the hierarchy of activity in the hierarchy of psychological phenomena. This correspondence can be revealed in different hierarchical structures, each hierarchical structure resulting in a respective stratification of the psychological science. For instance, let us recall the universal triad characterizing the structure of human activity:

$$O \to S \to P,$$

with the subject $S$ defined as a universal mediator producing things (products $P$) from the other things (objects $O$). On the most general level, the object, the subject and the product are the aspects of the same world:

$$nature \to spirit \to culture.[123]$$

Thus the spirit is defined as the most general form of subjectivity assuming all the possible special forms. Projecting this triad onto an individual subject (a person, a group, a society) as the agent of activity, we obtain the three basic representations of subjectivity in that agent:

$$consciousness \to self\text{-}consciousness \to reason.$$

Each of thus defined levels refers to a specific aspect of the agent, and its complete psychological characterization must synthesize the three mutually opposite and complementary characteristics.

---

[122] K. Marx characterized Hegel's philosophy as materialism "turned inside out".

[123] According to diathetical logic, the spirit as the mediating entity must be represented both in nature and in culture (the "second", consciously transformed nature). The projection of the spirit onto nature is what we call *psyche*, and this is the domain of psychological study. In culture, the spirit is represented in two-fold way: objectively, in the social institutes (corresponding to Hegel's "objective spirit"), and subjectively, in the cultural forms of reflection (which Hegel called "absolute spirit").

First, we consider the *individuality* of the agent, the features that make it an integral formation distinct from other agents (of course, taken from the psychological viewpoint, rather than mere physical separation). The corresponding level of science could be called *general psychology*, meaning that the presence of consciousness is the necessary prerequisite for self-consciousness and reason. An individual subject involved in a number of activities is the starting point for any psychological study. General psychology deals with motivation, typology, psychodynamics and temperament.

The second level, where consciousness becomes directed to the subject rather than the product of activity, is related to the psychological notion of *personality*. Here, the basic principles of general psychology are applied to a specific class of activities serving as an interface between the individual and the social environment (self-conscious behavior, the projection of social positions onto individual dispositions). The name of *differential psychology* seems to be appropriate for this level.[124]

Finally, an agent of reason is described in its relation to the product (or to the culture as the universal product). This kind of study brings knowledge about one's *mentality*, which is determined by one's place in the world, and one's place in the society first of all. This is the level of *social psychology*, considering social roles, psychological scripts and other phenomena of essentially collective origin.

In conformance with the basic triad $O \to S \to P$, general psychology is mainly concerned with the objective aspect of the subject's inner world; on this level, the correspondence between the forms of outer and inner activity is established. Psychology of personality studies the subject as the agent of activity, its ideal component. Finally, social psychology is to explore the productive aspects of the subject, the way of conscious beings consciously producing their consciousness.

All the three levels of psychological study are applicable to individual people as well to groups and societies.

In addition to these three levels, the psychological phenomena in non-conscious life is specifically studied in animal psychology, while the psychological components of the cultural level will probably be considered in a kind of cultural psychology.

---

[124] Traditionally, only the study of personal traits is referred to as differential psychology. In the hierarchical approach, however, the hierarchy of personality reflects the hierarchy of the person's activities; every such hierarchy could be interpreted as a personal trait.

## General Psychology

General psychology deals with the most common organization of activity, and its ideas are widely applicable in other fields of research. In a few words, general psychology could be characterized as a science about motivated behavior.

In Western psychology, motivation is commonly attributed to personality. However, the former Soviet psychological school considers motivation as a basic level common to all areas of psychology, reflecting the cultural status of the individual or a collective subject. Since motives can be personality-colored, studying motivation structures can bring knowledge of personality, but one's motives are not necessarily personal.

Historically, there were numerous attempts (*e.g.* in some branches of psychoanalysis, or in behaviorism) to reduce motivation is to organic impulses. This approach does not account for the social background of motivation, which is always present in conscious activity; organic processes can *resemble* motives, but, to become motives, they need to refer to a general need associated with a common activity, and hence to become subjectively mediated. That is why similar physiological states may induce quite different motivation structures. In the most primitive cases, motives can apparently originate from organic needs; however, these needs never form the core of the motive, which is essentially social. For instance, if I need food, or shelter, I do not need them merely to support my physiological existence—my primary purpose is to maintain all the cultural relations I am engaged to as a biological body; my death would cause problems to other human beings, which I would like to avoid as long as possible. Similarly, apparently biological desire of a longer life is socially transformed into a mechanism of cultural inheritance.

The hierarchical organization of activity as cyclic reproduction of both the object and the subject in their mutual interaction and reflection has been extensively studied in psychology, with numerous practical applications in psychological rehabilitation, psychotherapy, education, applied psychology, diagnostics and consulting. The fundamental ideas of the psychological theory of activity were developed in the former USSR since 1920's. Traditionally, the "culturological" approach by Lev Vygotsky was opposed to activity-oriented psychology (S. L. Rubinstein, A. N. Leontiev). However, the two schools become naturally synthesized within the hierarchical approach, representing the two positions of the same hierarchy.

## *Extent hierarchy*

In the psychological theory of activity, three basic levels are distinguished in the hierarchy of human behavior, namely, operation, action and activity.[125] Some relations between these levels are schematically shown in figure 4.

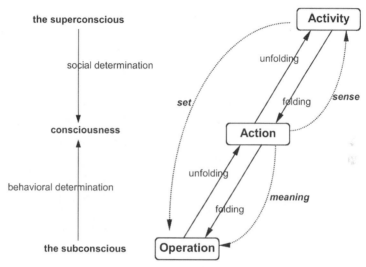

Figure 4. The hierarchy of activity.

As indicated in the scheme, both behavioral and social determination are important for consciousness. A person is engaged in some social activity, implying certain conscious actions; the actions are constructed by the person using the currently available operations that are performed in an automated manner and do not require conscious control. The sense of an action is determined by its role in some activity, while the meaning of an action indicates how it can be operationally implemented. Actions can be folded into operations in routine situations, or unfolded into activities, when new activity paths are to be sought.

Unfolding an activity in a sequence of actions (actualization) is not unique, and different sequences of actions can actualize the same activity. To say it more specifically, an activity is a *hierarchy* of actions, which can be unfolded in many ways. Similarly, the implementation of an action in a sequence of operations can also be achieved differently, obeying the

---

[125] A. N. Leontiev *Activity, consciousness and personality* (Englewood Cliffs, NJ: Prentice Hall Inc., 1978)

constraints imposed by the embedding activity; an action is thus understood as a hierarchy of operations. In this hierarchical structure, an operation is understood as the most elementary (folded) kind of a conscious act; attempts to unfold an operation result in an overall reorganization of the hierarchy, with the formation of a new operation level and the former actions unfolding themselves in full-scale activities.

The distinction between the activity extent levels can be illustrated by a quantitative analogy. Thus, an operation is subjectively performed in no time, and corresponds to the abstraction of a point, representing the discrete side of human behavior; on the other hand, an activity represents behavioral continuity, being essentially a process with a definite direction but no marked beginning or end. An action is the unity of continuity and discreteness: it is limited in time (one can complete an action and get a result), however, it is not dimensionless and syncretic like operations; the idea of a segment of a real axis comes to mind. Operations can be thought of as single moments of the pure (subjectively infinite) duration represented by an activity, while an action occupies an intermediate position between these extremes, spreading in time from the beginning to the end.

For integrity, human behavior must contain all the three levels. It is in their interaction that the phenomenon of consciousness is formed. Thus, an operation is too folded to admit any internal structure that could reflect the cultural side of consciousness. On the other side, an activity is too diffuse to achieve the characteristic specificity of conscious forms. One comes to the conclusion that consciousness must be an attribute of an action, with the levels of its operations and the embracing activity constituting the realm of the subconscious and superconscious, respectively. That is, for subjective integrity, any conscious action must be accompanied by the two opposite kinds of the unconscious, mediating their connection, like the subject in general mediates connections between objects.

The unconscious is not absolutely opposite to consciousness. Thus the level of the subconscious provides the behavioral foundation for consciousness, while the superconscious is nothing but another expression of consciousness' sociality. For instance, neural processes controlling the motion of one's hand are not conscious on themselves (subconscious), but the movement of the hand may well be aimed at a definite goal and be conscious. On the other side, one may be unaware of the motives of that movement as long as one's attention is focused on its goal rather than the embracing activity.

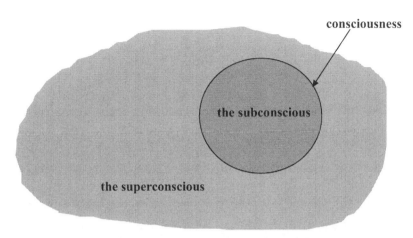

Figure 5. Consciousness as the boundary of the subject.

One more argument for the principal uniformity of operations, actions and activities comes from hierarchical refolding. Thus, any hierarchy can be folded an unfolded in different ways; in particular, activities can fold into actions, and actions fold into operations; also, operations and actions can unfold into actions and activities respectively (A. N. Leontiev). An activity becomes an action when it becomes a part of another activity, thus losing its self-motion. On the other hand, an action may become self-contained enough to develop into an activity. Similarly, a frequently repeated action may achieve a level of automation enough to drift off consciousness and become a subconsciously performed operation. Conversely, when a person's attention turns to the way of performing an operation (as in the case of an obstacle encountered), the operation becomes an action, which may sometimes cause a change in activity. Hence, the subconscious may be called the folded consciousness, and the superconscious may be called the subjective environment of consciousness, the zone of imminent development (L. Vygotsky). While consciousness is associated with the focus of awareness, the superconscious may be pictured as the rest of the field, not covered by the focal spot—the subconscious will then be the interior of the spot (figure 5). The subconscious is thus pictured as a part of some cultural space that has been immersed in the subject, interiorized. The immediate cultural environment of the subject forms the superconscious layer. Consciousness is the interface between the two, the mechanism of personal expansion in a given culture.

In figure 5, consciousness is represented by the thin line delimiting the subconscious from the superconscious, the boundary between them. The refolding of the subject's hierarchy (or the subject's development) could then be modeled by the changes in the "shape" of the subject—the boundary may get shrunk, expanded, or distorted, so that some areas of the subject's "exterior" would become its "inside," while some "inner" points would merge with the environment. Of course, this scheme does not refer to the physical implementation of the subject and has nothing to do with physical bodies or organisms, being related to them only in very special cases. It is the cultural space that is implied or, rather, the inner representation of the culture in the subject. In such an inner space, other subjects are represented by separate regions; interaction of such compact formations provides a model of personality and social dynamics.

## *Meaning and sense*

The meaning of an action is the hierarchy of operations that can become its means, while its sense is determined by the action's the position in the embracing activity (see figure 4). Such an understanding provides the basis for the study of meaning and sense in different sciences (psychology, linguistics *etc*). It is consistent with intuition: when one is asked about the meaning of something, this is usually a request for explication, unfolding the action implied into a specific operational structure; on the contrary, the questions about the sense of something imply a reference to the external circumstances reflected in the action, its "supreme" justification.

The convertibility of the hierarchy of activity leads to a kind of the relativity of meaning and sense, when apparently the same act may serve as either an element of some action's meaning, or the determinant of its sense. The natural corollary is that the same action may have quite different sense being considered in a different social environment, and one can never judge the other's actions without knowing their cultural roots.

Since any hierarchy can be unfolded in different hierarchical structures, the same action may be implemented by quite different means—however, this does not change its meaning, which is associated with the whole range of possible implementations within a given culture. The meaning of an action hence is dependent on the historically formed collection of possible operations, which, in a specific context, may appear as actions or activities. With the development of the society, the meaning of people's actions may drastically change, which should be accounted for when treating the subjective content of the events of the past.[126]

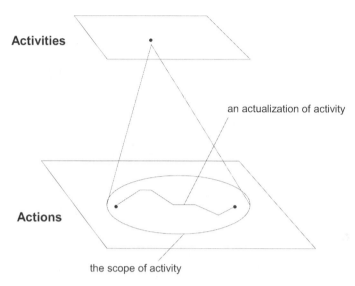

Figure 6. The scope of activity as an analog of meaning.

Actualization of activity is analogous to implementation of an action in a sequence of operations. The collection of possible representations of an activity by action sequences is reflected in the category of the activity's scope (figure 6). Indeed, the questions about the scope of an activity imply the specification of the actions that may serve to actualize it, and the possible "behavioral trajectories" actualizing the activity may only lie within its scope. Like the implementation of an action implies choice from a definite operation range (the meaning of the action), an activity is actualized in quite definite actions that are somehow compatible with it, belonging to its scope.

Any activity is a hierarchy of actions falling within its scope; conversely, each action is a stage of some activity, having sense only as an element of its hierarchy. An action can refer to many activities, so that its sense is also a hierarchy. The activity on the top of this hierarchy determines the awareness of the motives, providing a motivation for the action. In the same manner, an operation can serve as the means for different actions within different activities. A special process of rationalization selects one of them to

---

[126] For instance, Ancient Greek mythology, originally perceived as a kind of cosmogony, has become a sheer metaphor already by II century B.C., and it is used mainly as a source of abstract characters today. Many cultural traditions (like hand shaking, wedding rites, carnivals *etc*) go back to some actions that were quite meaningful centuries ago.

consciously justify the operation.

## *Psychological sets*

Since any hierarchical structure can be folded, activities can influence one's operations in an apparently direct manner (figure 4), which is known in psychology as set formation (D. N. Uznadze). This process resembles the already discussed folding of the primary sequence $S \to C \to R$ (a sensory input $S$ transformed into the internal state of the subject $C$, and then into reaction $R$) into a cognitive operation $S \Rightarrow R$ (coding of the image of the world $S$ with the operation patterns $R$), though the direction of influence is reversed. Obviously, removing the action as a behavioral link between operations and activities will push consciousness into the background, and people are often unaware of their sets.

Actions may have different sense when included in different activities—but the meaning of an action is also subject to influence of the embracing activity, which serves as a kind of filter selecting specific operations to use in implementing any action. Hence, one could distinguish the abstract meaning of an action determined by its general cultural compatibility with certain operations and its specific meaning within a particular activity (a specific position of hierarchy). This is a basic mechanism of the mutability of meaning: enhancement of the action's meaning by new operations and operational schemes can only occur through transferring schemes from one activity to another.

Beside "clipping" the action's meaning, activity can modify the formal properties of operations and the relations between them. Also, the outcome of an operation must be subjectively represented, with the available forms determined by the current activity.

The quasi-direct influence of activity on operations implies a hierarchy of activities, and different positions of that hierarchy result in numerous types of the set. In correspondence with the three kinds of inner structures, as expressed by the scheme $S \to C \to R$, there are sensory, perceptive and motor sets. In a sensory set, a person's sensations are filtered so that the external signal becomes coded by the motor schemes involved in the current activity, which may lead to illusions in perception. Similarly, in a perceptive set, the interpretation of sensations depends on the current forms of the subject's inner activity, which may result in an inadequate reaction to external stimuli. Finally, the current activity implies the subject's disposition to specific reactions, which may interfere with the conscious goals of the person, producing all kinds of mistakes.

Of course, sensory, perceptive and motor sets do not necessarily refer to physiology; the inner structures of the subject can be of a highly mediated nature, representing one's communicational positions, or self-reflection. On some other levels of subjectivity, the same triad of sets manifests itself as preference, mood or attitude. The mechanism of set formation lies in the basis of prejudice, obsessions, fanaticism, xenophobia *etc.*

The convertibility of hierarchies may lead to a rearrangement of the relations between their levels, and the indirect links responsible for set formation may become observable behavior. This requires "materialization" of the person's inner structures in an outer product, and thus communicating them to another person. Such techniques are widely used in psychotherapy to reveal sets and control them.

## *Psychological dimensions of activity*

The relations between a number of categories related to the hierarchy of activity in general psychology are illustrated in figure 7. On each extent level, for operations, actions or activities, the objective and subjective aspects in both reflection of the world (including self-reflection) and the subject's productivity can be distinguished; this leads to a two-dimensional model of human behavior within the level. This is a simplified version of the fully reflective description, which would account for the cyclic reproduction of activity in the sequence

$$O \to (S \to C \to R) \to P,$$

with all the possible lift-ups. Such approximate schemes are useful in practical applications, as long as one remains within the range of the model's applicability.

At the topmost level, the objective pre-requisites of an activity are introduced through the term "circumstances"; the subjective (ideal) side of the same external influence is called a motive. Motives are as external to the subject as circumstances,[127] but, since they objectify the subject's relation to other subjects (including the society as a collective subject), they are

---

[127] Motives could be called culturally mediated circumstances, hence referring to the product of activity $O \to S \to P$ rather than its object. This product-to-object

reflected in the subject as if they were originating from the inside; this is a typical illusion of self-perception. Similarly, the objective outcome of activity is referred to by the term "consequences",[128] while the subjective component of the same outcome is called a purpose, and it may seem independent of any objective consequences in reflection.

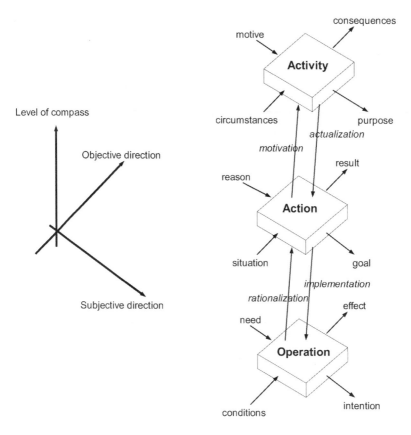

Figure 7. The dimensions of activity.

Like an activity is driven by some motive and following some purpose, an action has its specific reason and is directed to a definite goal; these are, along with the idea of the self, the main components of the conscious

---

[128] There is no activity without consequences (a product), however subtle the influence of the subject onto the world may be. Unlike the result of an action, the consequences of activity are too generalized to be consciously represented, but they are definitely sensible on the super-conscious level (cultural feedback).

representation of activity. Their objective aspect is described by the notions of the situation and the action's result. That is, what subjectively looks like achieving the goal is merely obtaining an objective result required by the current situation. While any action actualizes an embedding activity, there is a reverse process of associating an action with an activity, which is called motivation. The same action may belong to many quite different activities, which sometimes leads to a motivation conflict, which reflects the contradictions inherent in the circumstances of activity.

Folding an activity into an action is described by the notion of habit. When the activity becomes common enough, it may lose its motivating power, being continued just for conformity with the cultural requirements. This leads to less personal involvement and the loss of continuity, and the subject can perceive it as a whole, while retaining the feeling of duration. Thus the activity transforms into a conscious action. Conversely, when an action unfolds into an activity, the former goal does not matter ever more, and the circumstances are to provide a motive for the new activity.

On the level of operation, the objective dimension describes the transition from the conditions in which the operation is to be performed (the operational context) and the operation's effect; in the subjective dimension, an operation satisfies some need and is intended to produce a definite change in the world. Depending on the situation, an action can unfold itself into different sequences of operations—this is how the action is implemented; in this sense, operations provide the means of implementing an action. Conversely, one explains an operation relating it to some action, which is called rationalization. The place of an operation in the hierarchy of the action defines its function.

Practicing an action repeatedly folds it into an operation, since mechanically doing something does not require concentration on the process of achieving the goal, and only the awareness of the result is retained, the very process of operation becoming subconscious. Conversely, an operation can be unfolded into an action, so that the former embedding action may either become an operation, with a different actualization of the same activity, or it may remain a higher level relative to the former operation, thus becoming its embedding activity; this will require a corresponding restructuring of motivation.

Psychological models of any kind of activity must account for the nonlinear (reflective) character of mental processes. If we assume that any operation is a folded activity, the implication is that there are many latent acts of mental reflection, so that an operation is multiply repeated inside the mind before it comes to actual performance. Thus, an angle can be internally represented by the operation of building the angle of a definite size, which implies multiple repetitions and producing a sequence of angles constituting a kind of "overtone series" for the angle.[129] The usage of Fourier transforms (or momentum representations) is likely to be efficient in psychodynamics.

Activities never start just because the subject has some purpose, as well as actions are never initiated by a subjective goal, and operations are not triggered by an intention. On the contrary, the subject's purposes, goals and intentions originate from some natural or cultural processes. It is the cyclic character of any activity that provokes the illusion of behavior entirely determined by the subject's will. The objective development of culture produces the circumstances demanding certain activity; the activity itself is objective too, and it is bound to cause some objective changes in the world—in the subject, the objective determination of the activity is reflected as its motive, while the overall directedness of activity becomes a purpose. The purpose is always generalized, only determining a class of acceptable outcomes (products of activity). On the opposite, intentions are specific, being associated with a quite definite effect, which is what the operation is performed for. As expected, a goal is more specific than a purpose, but less specific than an intention; unlike the purpose and the intention, the goal is subjectively placed outside the subject, still being the subject's goal. The triad *intention → goal → purpose* pictures the goal as the dialectical negation of intention, and purpose as the negation of negation, reproducing certain aspects of its syncretic prototype. In other words, intention is understood as syncretic version of purpose, and purpose is regarded as socially mediated intention. Consciousness requires both material and ideal reproduction in both inner and outer activity; it is associated with the level of action. For the subject, intentions are determined by the conscious goal, while the purpose is only a premise for any conscious goals.

The scheme in figure 7 shows only a part of the existing relations. Thus, for instance, a motive can be understood as a mediated goal, while an objectivated purpose may provide a reason for an action. Description of the refoldings of the hierarchy requires considering interaction of different activities causing "shifts of consciousness" from one goal to another (A. N. Leontiev).

---

[129] P. B. Ivanov "A Hierarchical Theory of Aesthetic Perception: Scales in the Visual Arts" *Leonardo Music Journal*, **5**, 49–55 (1995)

## *Schemes of activity*

The hierarchical organization of every phenomenon implies a hierarchy of its manifestations. In particular, human activities are culturally related to each other, forming a hierarchy of qualitatively different activity types.

The ability of making one's own actions a part of external stimulation, as described by the link $R \to S$, is what distinguishes an individual as a subject from mere animal. That is, any stimulus $S$ is a hierarchy, containing both low level stimulation, simple reflection of the world, and a subjective component, marking the external stimuli as *produced* by some subject (of this or a different level). For an animal, the stimulus is only an imprint of the current situation, carrying the same biological sense regardless of how it has been produced. For conscious humans, any situation is primarily important by its social consequences, and not its physiological meaning. This results in a different processing of the stimuli in an individual subject, as compared to the animal. In particular, each stimulus does not influence the subject's inner activities on itself, but rather through a number of pre-established filters, which perform primary categorization and reveal the subject component. Such subjective stimuli are known as *sensations*.

For an individual subject, a hierarchy of reflexes becomes folded in an integral *sensation*, which is transformed into a certain *representation* of the world. The basic scheme of the inner organization of the subject

$$S \to C \to R$$

is then readily reinterpreted as the well known cognitive triad:

*sensation $\to$ perception $\to$ representation.*

According to the principles of hierarchical logic, each category in the schemes can represent both a static (structural) aspect of the whole, and its systemic side (process). Therefore, "sensation" here means both the process of sensing, and its results, a mental structure.

Since hierarchy can be unfolded differently, the same activity can appear on different levels, depending on the social circumstances and the presence of other activities. For instance, in the folded form, sensation is represented by an inner structure $S$, which is transformed into a perceptive structure $C$, which, in its turn, initiates folded (latent) reactions $R$ as an internal representation of the situation: $S \to C \to R$. In this case $S$, $C$ and $R$ denote mere operations within some other activity. However, sensation, perception and representation can also develop into conscious actions, or full-scale activities; in the latter case, they must have appropriate motives and hence be socially mediated. Thus, the activity of *listening* folds into the action of *attending* focusing awareness on certain aspects of the whole sound pattern; in the most folded form, it becomes the operation of *hearing*, immediate awareness of the sound. However, listening does not mean full consciousness: one can listen, but hear nothing; similarly, hearing lacks full consciousness too, since one can hear something without realizing what is heard. In special situations, hearing may unfold itself into a conscious act, and even transform into listening.[130] Similarly, in vision, the activity of *viewing* (or observation) can be folded into the action of *looking*, which, in its turn, folds into the operation of *seeing*; the unfolding of seeing into looking or viewing is governed by the same laws as for hearing.

In these examples, the hierarchies *hearing → attending → listening*, or *seeing → looking → viewing* are not immediately related to the triad

*sensation → perception → representation*;

rather, they represent another dimension of hierarchy. Listening can develop on the sensory level, and audile representations exist in hearing. The order of levels is relative in any hierarchy, and what was an activity in one position of the hierarchy, can become an operation in another position. Thus, the level of listening cannot be said to always lie above the level of hearing, because perceptive hearing is definitely superior to sensory listening—though both refer to physiologically the same process. The synthesis of hierarchical folding and unfolding is followed by the act of "merging" the opposites in a higher-level integrity, according to the general logical scheme: *analysis → synthesis → lift-up (Aufhebung)*. Consequently, there are two opposite ways of converting sensory listening to hearing, via folding and lift-up, producing either a sensory operation or a perceptive operation.

Another instance of the hierarchical relations between activities is provided by the development of the *same* activity increasing the degree of social mediation. Thus, there are different levels of perception, from the primitive animal forms to the sublime spirituality of aesthetic, logical or ethical schemes.

Activities taken as a whole differ by generality, forming an objective hierarchy of activities.[131] The convertibility of hierarchies moves the same

---

[130] P. B. Ivanov "A Hierarchical Theory of Aesthetic Perception: Scales in the Visual Arts" *Leonardo Music Journal*, **5**, 49–55 (1995)

[131] The subjective projections of that hierarchy constitute one's personality.

activity to different levels, depending on the social conditions and the presence of other activities; however, there is the universal direction from inside the subject to the social environment. Thus, one can also consider sensation, perception and representation as the three consecutive stages of a higher-level activity of *orientation*. In its turn, lifted up orientation becomes an analog of sensation in triad of activities of yet higher level:

$$orientation \to comprehension \to realization.$$

However, within other activities, this triad too can represent either a sequence of operations within an action, or a sequence of actions in an activity, or hierarchical ordering of separate activities. The sub-levels analogous to sensation, perception and representation within orientation can also be distinguished within comprehension and realization. On the topmost level, one finds a very general version of the same scheme:[132]

$$consumption \to assimilation \to production;$$

obviously, these are the three stages of any activity at all, which can be associated with the arrows in the scheme

$$O \to \underset{\curvearrowleft}{S} \to P$$

That is, any object $O$ is first to be consumed by the subject $S$, then a series of internal transformations (inner activity) is initiated in the subject, which eventually results in producing an outer effect, the product of activity $P$.

Obviously, sensation can be considered as a special case of consumption, and representation is a special case of production, with the product being a subjective structure. This indicates that the above hierarchy of activities forms a complete triad. The levels of this hierarchy correspond to the levels of the subject that can be denoted as individuality, personality and mentality. Since the same psychological dimensions are present in any activity of any level, one can consider individual, personal and social motives, needs, or goals, *etc*. However such distinctions are never rigid, since conversion of hierarchy can exteriorize an internal motive and interiorize external motives; this conversion has little to do with awareness, which only refers to the levels of extent, the distinction of activities, actions and operations.

A special case of hierarchical convertibility is provided by inner conversion, projecting one dimension of hierarchy onto another. There is a generic process of the transformation of the hierarchy of activities into the

---

[132] This is yet another interpretation of the formula $S \to C \to R$.

organization of a single activity, and back. Primarily, in the triad

$$sensation \to perception \to representation,$$

the components are separate (inner or outer) activities, either coexisting or sequential. In a folded form, only one of them will represent an activity, with the other two playing the role of actions and operations. Thus, sensation, perception and representation can become the levels of the extent hierarchy (figure 7); in this context, one would deal with conscious perception, subconscious sensation and socially determined representations. Upon the next folding, sensation, perception and representation would become the three classes of actions that actualize an orientation activity. Folding the hierarchy yet another time, one comes to sensory, perceptive and motor operations as the means of conscious orientation within a more generalized activity.

The universal character of subjectivity implies universality of any activity; in particular, the subjects of any kind must have similar activities and develop through similar stages. The objective side of this universality goes back to the general laws of the world's development, including development of the subject as a part of nature. Inner development of the subject results in the universal forms of spirituality. In the objectified form, as a cultural product, this universality is reflected in the schemes of activity.

Every culture has its own schemes of activity, and communication between different cultures is only possible through shared activity schemes. However, the very idea of subjectivity implies that there will always exist some points of contact, and every two forms of reason can understand each other; moreover, mutual understanding of different cultures is bound to grow with more contacts, with the difference in the scope of their subjectivity gradually diminishing.

The collection of schemes built in the culture provides an important mechanism of maintaining the integrity of the subject on every level. The communicational integrity is achieved through the commonality of the standard forms of communication; this implies the existence of standard personal roles, and hence the standard components of the personality, which then become projected onto every psychic process. Thus, the standard ways of dealing with objects one encounters in everyday life get built into the standard ways of perception.[133] This is different from the formation of the perceptive set in that the culturally supplied perception forms are shared by

---

[133] This is often used in the arts to efficiently control the audience's attention.

many people and they do not vanish after acquiring more behavioral patterns.[134]


## Differential psychology

There have been numerous attempts to discover the true structure of personality, once and forever. But the answer escapes, and it is almost obvious why. Psychology still lacks the very definition of personality, always referring to something intuitively felt as a common core of apparently variable behavior, but never clearly known or communicated. That is why some researchers may try to characterize personality in motivational terms, while others will borrow their notions from social psychology, or even from outside psychology as such.

On the other hand, personality can be regarded from many angles, and there is no agreement about the choice of the fundamental dimensions. Is there any significant structuring at all?

I suppose that the hierarchical approach provides an acceptable answer. Personality is not a structure, nor can it be reduced to a system. As a hierarchy, it can be unfolded in different ways, manifesting numerous structures and systems in different aspects. However, such structures are never arbitrary; they must obey the general laws of hierarchical development. In other words, there is no preferable structure in personality, but rather a unifying principle of their construction.

To develop unified psychology of personality, one needs to move from consciousness to self-consciousness, considering self-reflection as a special kind of activity. But any self-reflection is only possible through comparison with the others, and hence communication. On this level, we pay attention to how people do something rather that to what they do. Personality manifests itself in any activity at all as a peculiar mode of behavior. People can produce the same thing differently, and the way they act is a product too. This subjective product can be formally considered as the product of separate activity; in differential psychology this aspect of activity is the center of subordination for all the other activities, and personality becomes defined through the formation of motivational hierarchies.

Since behavioral manifestations of consciousness are innumerable, one

---

[134] Rather, psychological sets are constructed from the standard cultural forms.

could doubt whether it is possible to scientifically study subjectivity at all. Every person seems to experience the world in a unique way, using highly individual and incommunicable inner representations. One can never feel exactly like another, or think like another—and even in science, the reported results can be differently understood by different people of different epochs. There seems to be no categories applicable to everybody, and no universal laws governing any personality at all. This conceptual indeterminacy can hinder the development of scientific methods in psychology, replacing it with vague belletristic descriptions, or philosophical speculations.

But it is not enough to merely say that somebody's behavior is guided by "individual constructs"; the origin and development of such constructs has to be primarily explained, as well as the possible ways of their practical usage and control. However, as soon as the origin of anything is to be considered, there is already comparison and hence commonality, allowing for analysis and categorization. Any activity at all implies categorization, and science is no exception. Bringing people under various categories is yet another side of individuality: like individuality becomes a universal mechanism of activity, the historically formed general schemes of activity provide a framework for studying the specificity of personality. A psychologist *must* discover commonality in people and treat them as abstract categories—otherwise, there would be no science at all. Still, a psychologist must also clearly understand that any special case always differs from the general rule it represents; to describe the character of difference, new general categories are to be developed—thus knowledge becomes hierarchical, and we come to the understanding of specificity as a hierarchy of general traits.

Categories used by one psychologist are certain to differ from those used by another psychologist, even belonging to the same scientific school. Quite often, the categorization chosen by a psychologist to interpret one's behavior has little in common with one's self-perception and self-identification. Different (and not necessarily complementing each other) approaches can be equally valid, representing the different positions of the same hierarchy. As long as mere interpretation of behavior is concerned, neither model is preferable. The preferential position of hierarchy can only come forth under certain external conditions, in a particular application. Thus, a personal profile constructed using some psychological test will not be usable for prescribing any therapeutic procedures, unless the organization of the test reflects the essential features of the cultural environment and the typical modes of activity. In any case, selection of a particular psychological model depends on the personal problem to resolve.[135]

Some psychologists (such as H. Eysenck) advocated entirely differential study, with the individual considered as a point in a multidimensional space of psychological parameters. The opposite trend was promoted by G. Allport, who insisted that every person was absolutely individual, and psychology was only to describe one's uniqueness. From the hierarchical viewpoint, either stand is insufficient. Science is not concerned with individuality; its main purpose is to discover regularities. In a scientific study, a person will necessarily be placed in a social framework and described by the measure of belonging to the community. However, there are many social scales, as many as conscious activities; therefore, one will always observe a hierarchy of scales that is unique for each particular person. An individual is the topmost element and representative of a specific position of the hierarchy.

Practical needs limit the scope of science. There is no use of merely pondering on somebody's uniqueness. On the other hand, people don't want sheer abstractions equally applicable to everybody. We need effective tools for operating on our own personality, and we expect them from science, the more the better. But no science will tell us what must be done with a particular person in some special circumstances. The choice of instruments is influences by the cultural conditions and their reflection in personal preferences. And the way people use the available tools distinguishes one person from another.

This is like a doctor prescribing a different treatment for each particular case of disease. Doctors do not need to invent medicines individually for each patient, and each medicine can be used by many very different people. Doctors only combine the existing medical techniques to produce the desired effect. Theoretically, it may be desirable to be as specific as possible, but absolute individuality of treatment is hardly ever attainable.[136] Even if we learn to operate on the molecular level, the hypothetic nanotechnologies will still be the same, with only their application differing from one case to another. And, of course, we still have to account for the social conditions that essentially influence human physiology.

---

[135] There is analogy with quantum mechanics, where all the complete basis sets are equivalent, but some calculations become much simpler in a definite basis, being intractable in any other. Similarly, many field theory problems can be solved in a preferential gauge only, which demands checking the applicability of the results to other gauges.

[136] Similarly, engineering is a practical discipline that knows how to transform the scientifically discovered phenomena in real devices, combining the available techniques in a specific way.

## *Hierarchy of personality*

Personality presents conscious activity in a reflected form. Consequently, the hierarchy of personality will reproduce the hierarchy of activity described in the previous section (figure 4), but this hierarchy will be presented in a reflected form (a different position).

Studying personality, we start from an activity as an elementary unit. However, activities are considered here as partial manifestations of the subject's inner core (personal traits). This core is rather stable, and it makes all the subject's activities resemble each other, being marked with a common personal style.

On the next level, we find behavioral act as hierarchies of activities, assuming the corresponding motivational hierarchies as its inner representation.[137] A behavioral act is directly related to self-conscious choice, to making decisions. While conscious actions are performed for some reasons and directed to certain goals, behavioral acts grow from personal dispositions and moods. Here, the subject is essentially *the self*.

On the highest level, one deals with the integral behavior as a hierarchy of behavioral acts. The personal mode of behavior is determined by the integrity of the person's character as a hierarchy of traits.

Since any hierarchy can present itself in many ways, unfolding various hierarchical structures, one will observe different structures in personality under different view angles. However, all these conversions conserve a kind of integrity that may seem to allude to the presence of some inner agent as an active substance controlling one's inner and outer behavior. This makes psychologists personalize the component of one's psyche under the names of *Ego*, *Id*, or *Self*.[138] Normally, such personalization is only an attractive metaphor: there are no inner agents in personality; on the contrary, it is the person as a whole who acts, and the person's activity obeys some general laws, which result in a definite inner organization. This psychological organization is not built-in, inherent, or genetically pre-determined. Rather, all we call individual, or personal, is merely selection of objective pre-dispositions by the social imperatives.

In particular, personal traits do not arise from inside the person. They are

---

[137] There is also a purpose hierarchy.

[138] Since a conscious agent is the necessary component of any activity, and personality is revealed in characteristic hierarchies of activities, one could picture personality as a hierarchy of individuals, the possible guises of the person.

imposed on an individual by the society through the cultural context and its gradual interiorization.

The development of personality (the growth if its hierarchy) always follows a certain objectively set pattern. There is a sequence of distinct stages, later becoming the levels of personality. However, this sequence is not rigid and it may differ for different people, since different aspects of the culture dominate in different social conditions (as determined by the cultural formation, the overall type of culture). Though psychological development is directed by the social environment. individual behavior can significantly vary, still remaining limited by a culturally determined range of activities; some choices are socially preferable, while the rest will become suppressed, or at least underdeveloped. If the social conditions drastically change, one may be driven to the necessity of as drastically changing one's psychology. This is a crisis in personality, a personal drama.

Recalling the definition of consciousness as the boundary of the subject, separating the domains of the subconscious and the superconscious (figure 5), and combining this picture with the idea of self-consciousness as consciousness within consciousness, one can conclude that the inner organization of personality is nothing but the inner complexity of the boundary; a thin line on the level of individuality becomes a broad (and well structured) band in differential psychology, the self. The development of personality implies changing the boundary of the self, as well its inner hierarchy. The bodily boundary of the individual can be stable during this process—it is only the links between the person and the society that change.

## Social Psychology

On the level of social psychology, the inner organization of the subject becomes explicated in the organization of a higher level subject, a group. Social psychology is different from group psychology or mass psychology; the latter deal with a collective subject as a whole that can be considered from the viewpoint of general, differential or social psychology. In social psychology, we are interested in the behavioral regularities and restrictions imposed by the interaction of the subject with a group.

The hierarchy of activity as pictured in figure 4 has its counterpart in social psychology as well. On the lowest level, one finds behavioral patterns, social *roles*. Different hierarchies of roles correspond to various *ways of life*.

Finally, on the highest level, there is a general directedness of the subject's existence that could be called *destiny*. The sense of one's life is thus defined as the position of the chosen way of life in the hierarchy of one's destiny. Similarly, one can consider social sets[139] *etc.*

The levels of this hierarchy can be projected onto any individual activity, making it a social event rather than mere objective process.

One's way of life is determined by one's social position, which dictates the choice of acceptable roles. People are governed by their interests, not instincts. Even the basic physiological needs can be suppressed if one's convictions and interests demand it.

Modern social psychology lacks clear understanding of life styles and cultural predestination, mainly concentrating on role behavior. Moreover, without considering to the higher levels of hierarchy, psychologists have to discuss the communicational aspects of activity, while the real source of any psychological phenomena is in production.

Roles can be categorized in many complementary ways. Three general communicative positions have been introduced by E. Berne:

$$child \to adult \to parent$$

or, formally,

$$C \to A \to P$$

The child position refers to the passive mode of activity, expecting guidance from outside (from the communication partner). The parent position means communication "from above", actively imposing one's ideas on the others. The necessary link between the two poles is represented by a well-balanced behavior of an adult.

In the hierarchical approach, this scheme can be extended, allowing for conversions of hierarchy, structural and systemic interpretations, cyclic reproduction and hierarchical development. This hierarchy is analogous to the universal triad $S \to C \to R$, with similar secondary form $P \Rightarrow A \Rightarrow C$ and various dyadic cycles.[140] This hierarchy of personal positions is essentially the result of interaction with the cultural environment, including both production and communication.

---

[139] The mechanisms of set formation can be used under certain social conditions to control people's behavior through eliminating conscious action, thus limiting the people's universality and preventing them from developing into true subjects.

[140] The possibility of such extension was implied by E. Berne's, as he considered such constructs as "parent in the adult", "child in the adult" *etc.*

## Transactional analysis

Treating communications as an activity, one could study its possible actualizations and their operational structure. The elementary communicative operations within a communicative action are known as *transactions*. Obviously, the intended effect of a transaction depends on the action it belongs to, and the motives of the embedding activity.

Depending on the involvement of personality in communication, three kinds of transactions could be distinguished. On the *syncretic level*, a person is not separated from the communication act itself; the person's participation in communication is random, determined by the person's character and the current mood. *Analytical* transactions involve conscious choice of position under the influence of some external reasons rather than personal sympathies or repugnance. On the *synthetic level*, conscious communication becomes an intrinsic demand, any private movements correlated to external necessity.

The transactions of the lowest level are closely related to the standard forms of behavior acquired through individual learning in a specific small-group environment (today, this is mostly the influence of the family). Such behavioral standards can be inherited form the reference group as transactional *scripts*, and this inheritance has much formal similarity to genetic inheritance in biology, with the transactional sets of the "psychological parents" taking the place of biological genome.

The analytical level of communication is characterized by the formation of formal schemes, which could be compared to a game with fixed rules adopted by all the participants by convention. Though there is a great number of such "games", they can usually be reduced to a smaller number of typical schemes, and the study of the transactional structure of communication becomes analytical indeed.[141] However, the basic transactional schemes strongly depend on the culture and the level of social development of the "players". The global commonality of elementary "games" throughout the world must be attributed to the universality of the activity and the objective laws of socio-economic development.

Communication on the synthetic level is the most difficult to grasp, since it does not allow detaching the observer from the communication act, and the possible experimenting is essentially introspective and reconstructive. This kind of communication could be called *intimacy*; this is the most "human" form of human contacts. Friendship and love are the evident examples of

---

[141] This is the case primarily described by E. Berne.

synthetic communication.

The distinction of these levels of involvement does not, in general, correlate with awareness. There is conscious syncretic communication (*e.g.* group entertainment)—and subconscious psychological games. Though syncretism would generally mask conscious behavior, and analyticity favors realization of one's communicational preferences, there is no simple parallel, since the formation of the hierarchy of communication is determined by communication context.

On the level of personality, involvement is determined "from above", by the person's position in a hierarchy of small and large groups, up to the society as a whole. Thus, informal groups support syncretic communication, while formal groups (*e.g.* the personnel of a company) demand a certain level of analyticity; intimacy is generally the property of a very special kind of groups—the "close-contact" groups (a company of friends *vs.* companions; lovers *vs.* a family; advocates of the same views and convictions *vs.* the members of a party).

However, there are no absolutely formal or purely informal groups, and every group combines different features, in a specific proportion. Therefore, intimacy can usually be observed as a superstructure of formal relations.[142] The type of the group is determined by the position of the hierarchy of "formality", always retaining all the levels[143]. The members of the group reflect this hierarchy in their own way, constructing their own hierarchical structures as the subjective models of the objectively unfolding relations. When these subjective structures do not coincide, conflicts arise.

Any transaction occurs on all the levels of the hierarchy of involvement, and the positions of the communicants differ on different levels. Since only one of the levels is associated with awareness (the communicative action a hierarchy of transactions), there are unconscious communicative positions; interaction between such unconscious positions results in various *hidden transactions*. Thus, the balance of syncretic and analytical transactions can lead to apparent contradiction between the "form" and "content" of communication[144].

---

[142] Psychological games cause real pain when some intimacy is involved.

[143] It would be easy to work in a company, where the staff could formally perform their duties (the ideal of an executive). There have been so many dreams about sublime love devoid of any nuisances of the real life... In reality, personal attitudes of the company's personnel strongly influence their performance, and there is no love without formal responsibility.

[144] For example, a syncretic transaction of the $C-P$ type can look like an $A-A$ transaction: two colleagues are discussing a business problem—but one of them tries to find

The difference of transaction schemes on different levels does not lead to conflicts on itself—it is the difference in the schemes of activity that matters. When the partners have similar conceptions of communication on all the levels, there is no behavioral contradiction.[145] However, if the partners differently identify the levels of hierarchy and use different involvement schemes, the outburst of a conflict is a matter of time. Thus, if one partner seeks for relaxation in formal communication, it is potentially dangerous to discuss global problems or talk confidentially. Quite often transaction conflicts result from confusion of informality with intimacy.

On the surface, the problem is in that one partner suggests the other a communicative position different from the position the other would like to take. However, such perceptive illusions and set conflicts are only the consequence of the external conditions of communication and a specific organization of the social environment—therefore, the only way to really cure them is to rearrange the group, and maybe something in the society producing such groups. As a rule, people adequately reflect their environment, and the adoption of one or another scheme of transaction means the existence of an objective premise for that.[146]

The structure of the group determines the possible types of inner communication. Since the same group can differently unfold its hierarchy, a kind of synchronization process for individual involvement. Transactional conflicts arise when the members of the group cannot synchronize conversion of their involvement hierarchies. When a group does not allow its members to correlate the hierarchical structures they produce in their minds, this indicates lack of hierarchy, merging all the levels in an amorphous conglomerate. Primarily, a transactional conflict is a mechanism of hierarchy development, the contradictions being resolve by driving the opposites to the different levels of hierarchy. In a conflicting group, the relations thus become more formal, and internal models get correlated by the higher level. Further discharge of the conflict assumes gradual correlation of all the levels, from top to bottom (externally mediated unfolding of hierarchy). The structures produced in this process depend on the social conditions rather than on individual will; the motivation structures of the members of the group (and

---

support in the other, while the other only wants to "publish" his opinion.

[145] For instance, they are playing the same psychological game, as described by E. Berne. There is a Russian proverb: "Милые бранятся — только тешатся" ("Sweethearts quarrel just for fun").

[146] For example, the position of the boss of a company implies transactions $P - A$ or $P - C$, while the employees of the same level are expected to use the $A - A$ transactions.

consequently, the structure of their consciousness) are objectively determined. Thus, if the existence of the group is socially adequate, its internal conflicts will produce new hierarchical structures; if, however, the group is socially deprecated, it tends to split into a number of struggling fractions (just another formal structure) and finally decay into several groups, with rapid social divergence.

There are many possible inner representations of the group that are related to different kinds of conflicts. The universal involvement hierarchy is a good model for testing the methods of enhancement of both group stability and individual self-determination.

Since analytical transactions are correlated with the formal aspect of the group, while the syncretic level includes all the informal transactions, the mechanisms of effectively joining the people in the community making the group a unity can only belong to the synthetic level. The important corollary is that no group can be stable if there is no synthetic involvement in it—that is, if there is no trace of intimacy in communication within the group, no love or friendship. This is not as trivial result as it may seem, since it implies the objective character of both antipathy and love, and the presence of a material basis leaves no room for the subjectivist appeals of the "love thy neighbor" type. One cannot just love or hate somebody else—there must be something to love for, or hate for. Accordingly, the unity of the group is in no way a result of mutual consent or convention—on the contrary, any consensus is a result of the group's unity achieved due to specific social and historical requirements.

The techniques of psychological manipulation through specially intended transactions are designed to work within the same involvement level: all the partners similarly treat communication as formal, or pastime, or allowing intimacy. However, the usage of such techniques will be different on different levels. Thus, there is no use trying to change the partner's position in syncretic communication, since this position can only drift by itself in the course of syncretic communication, through multiple transactions; the typical trick is to accept the position of the partner slightly exaggerating it to provoke the avoidance behavior. On the analytical level, the swing technique is widely applicable: recognize the partner's superiority to make the partner recognize your own merits (mutual "strokes"); this implies a damped periodic motion from the $C-P$ to $P-C$ transaction type, which gives the standard $A-A$ transaction in the average, gradually approaching it.

The synthetic level of communication is of special interest. It may seem that, if two people are really close to each other, there is no need in any psychological techniques at all, as the very intimacy of their communication apparently implies. However, this impression is deceiving. Intimacy is the synthesis, unity of syncretism and analyticity, the internal motion from one level to another and back, their mutual transformations. Consequently, the techniques of the syncretic and analytical levels must be applicable here in a "lifted up" form, in a mutable combination. There can be no intimacy without mutual "strokes"—however, such analytical transactions occur syncretically, in the form of mutual acceptance. Random syncretic transactions can become a favorite psychological game for sweethearts. This is the principle of intimacy: live playing, play seriously.

Similarly, though universality means freedom, conscious activity cannot be free from anything at all—rather, it involves transformation of external constraints into internal limitations, and back; restrictions are not merely removed, but the very freedom becomes a regulative principle implying a deliberate care for all the necessary limitations: self-limitation and spiritual freedom become the same.

Of course, such a synthesis is only possible under definite social conditions. In the society, where its members have to compete with each other to grasp at a portion of the public wealth, there are few opportunities for real intimacy. Intimacy demands rather high level of economic and social development, and every trace of the dialectic unity of external and internal regulative mechanisms in the people's behavior indicates some social progress.

## *Collective behavior*

When a subject is included in another subject as its part or element, the lower level subject's behavior becomes modified from the higher level. The same person (or group) simultaneously plays two opposite (and often contradictory) roles: primarily, an active individual and the agent of a hierarchy of activities—and also a representative of a collective subject and a part of some externally prescribed cultural motion. This may result in "split" personality, showing its different sides in different situations in a manner that can sometimes startle the friends and relatives. Indeed, any person has to pass through such a "split" phase in a transitory way, when included in a new group, until the inner hierarchies develop additional levels to assimilate the new social position.

The interplay of "intrinsic" and "imposed" motivation is the principal mechanism of personal development. The characteristic times of hierarchy rotation are of primary importance here. If the inner and outer conversion cycles have very different characteristic times, there is usually no problem in merging the two hierarchies in one. In, however, the periods of rotation are close to each other, collective and individual behavior will significantly interfere; the typical beats that arise in such interference manifest themselves as unmotivated deviatory behavior, unexpected reactions, reduced productivity *etc*. The subject cannot control such behavioral effects, since they are entirely due to hierarchical dynamics, objective rather than subjective. Similar interference can also be observed when a person has several types of collective behavior, being a member of several different groups at once. People are often accused of inadequate actions that do not indeed have any relation to consciousness and subjectivity; the true source of such "freaks" should be sought for in the people's social environment and competing obligations.

The dominance of collective behavior can sometimes be destructive for subjectivity. Consciousness assumes freedom, and the individual must periodically return to inner dynamics, forgetting about any external pressure. If induced motivation is always on the topmost level of motivational hierarchy, subjectivity degrades to mere reactivity more appropriate for an animal rather than a conscious being. In the case of panic, mass psychoses, stubborn superstitions, overwhelming trade enthusiasm, or warfare, people lose their universality, becoming the parts of some whole, objects rather than subjects.

## Animal psychology

Since a living thing has a soul, there is something to consider for a psychologist. However, the psychology of primitive animals is limited by reflexes, or emotions of the James-Lange type. Human interest to animal psychology comes from human needs; we seek there for explanations of the mechanisms of our own behavior. Psychology of higher animals is much more promising in that respect, especially when the animals communicate with humans.

The unity of the world implies the unity of the animal and social levels too. There is no absolute difference between humans and animals—and the subject can be represented in organic bodies only in a relative way.[147] That is why one can describe animal behavior in anthropomorphic terms, paying special attention to the applicability of the results obtained. Every phenomenon can be described on different levels, and all such descriptions will be of equal adequacy and value, since they refer to the different positions of the same hierarchy; behavior as such is different from its physiological implementations. The same behavioral act contains both social and physiological components, the former determining the organization of the latter via a long chain of mediations.

The guiding principle for comparing human and animal psychology is that the same behavior can be implemented by different means. Thus, both a human and a cat may equally feel sympathy and aversion, be afraid of something or show wonders of braveness; both may be clever or foolish, sly or open-hearted, shy or communicative... However, all those psychological characteristics will look entirely different on the physiological level, just because the structure of the human body is different from that of the cat's body, and this implies different physiology for apparently the same behavioral outcome. Even identical physiology does not imply identical implementation of behavioral schemes.[148]

Animals living with humans acquire the schemes of behavior non-typical of a wild animal of the same species; corresponding physiological mechanisms have to develop to support the new type of behavior that does not have any evolutional justification in the natural environment. Humans living with animals experience their behavioral influence as well, borrowing the schemes of activity from the animals. However, human physiology allows humans to borrow behavioral habits from any other animals, and it is the ability of combining quite different behavioral schemes that puts the *homo* family in a preferential position to develop subjectivity; this is one of the aspects of the subject's universality. Normally, an animal learns from the

---

[147] Thus, it can be found that the purely physiological explanations of the cat's behavior abounding in the literature can as well be applied to human behavior in similar situations, including the forms that most researchers would call specifically human (love, fantasies, dreams, sorrow *etc*). Instead of revealing the social roots of human-like behavior in cats, some psychologists chose to reduce humans to animals.

[148] This could be compared to designing a portable computer application to run on different platforms, which often requires adapting the high-level algorithms to the hardware used. Even within the same platform, one has to use special drivers (ports) to communicate with peripheral devices manufactured by different companies; low level implementation of the same algorithm is entirely dependent on the hardware. On the same computer, different operation systems differently control the same equipment.

animal of the same species (or its substitute, in the domestic case), while a human learns from all the living creatures (and even some inanimate things) together.

Higher animals are much organically flexible; they can communicate with humans and learn from them, and become socially educated. Of course, if one treats a kitten like an animal, making it live like an animal and behave like an animal, one cannot expect any human-like forms of behavior in the mature cat. But the same holds for humans brought up by the animals: they cannot properly adapt to the human society. We also know about humans losing their human character after the contacts with the rest of the humanity have been broken. Modern people are much like animals in many respects; quite often, people show examples of a very unreasonable sort to the animals.

If a kitten becomes a member of a human family, treated as sensibly as possible, much talked to and stimulated to behave in a civilized way—a cat grown up from this kitten will be not entirely animal; all the genetically determined instincts will have to transform to support a different type of behavior, and many instincts can be socially suppressed (up to the sexual sphere, where both cats and humans are very difficult to control).

# ASPECTS OF CONSCIOUSNESS

There would be no use in reflection upon consciousness and subjectivity, if it could not assist people in their productive activity, including creative reproduction of the subject on all the levels. There are as many applied aspects of the philosophy of consciousness as different activities within the cultural formation; it is hardly possible to enumerate and describe them all. However, the necessity of accounting for the subjective side of any cultural phenomenon (and psychological phenomena in particular) has long since become commonplace.

There are three general directions of conscious development of our subjectivity, as indicated by the universal structure of activity:

$$object \to subject \to product,$$

or, formally, $O \to S \to P$. First of all, the position of the subject in the universe must be comprehended, to more efficiently adjust the world to human needs and prevent disastrous mistakes; our understanding of the world will be more adequate with the subjective determinant of objective phenomena taken into account.[149] The next direction is related to increasing the power of self-regulation, making it comprehensive, universal like the subject as such. Deliberate reorganization of our own subjectivity is an unmatched challenge to our reason; animals can use things and manipulate with other creatures to satisfy their needs—but it is only conscious beings that can use people and things to change themselves. The productive aspect of the subject's self-development means consciously rearrangement of the hierarchy of production and culture, introducing new types of subject-object and inter-subject relations stimulating people's creativity. Finally, these three aspects will be synthesized in the hierarchical development of the subject.

---

[149] This is especially important in science, pretending to produce pure objectivity, but easily yielding to philosophical idealism and mysticism in case of difficulty.

## Consciousness and Physiology

Though the roots of subjectivity are in reflexive mediation on the physical and biological levels, not any material system can serve as a substrate of reason. Consciousness is a relation between material bodies, but there are relations not associated with subjectivity, and studying consciousness requires clear distinction of the subject from the non-conscious world.

The material implementation of the subject combines both organic and cultural elements, and this compound body is steadily extending through using tools and instruments, asymptotically encompassing the whole universe (K. Marx). Organic bodies participating in this process acquire specific properties and functions that mark them as "conscious" bodies.

External factors dominate in producing conscious behavior, and the demand of supporting as much universality as possible precedes considering any organic structures or functionality. There is no use seeking for neural correlates of conscious acts or activity structures; rather, one would first formulate the conditions necessary for a material system to be compatible with certain aspects of consciousness, and then seek for physiological mechanisms compliant with these specifications.

### *Mind and body*

Philosophical idealism does not much care about relations of the mind to any material body; most people cannot accept such an unconstructive attitude. The obvious connectedness of our consciousness with our biological body demands explanation; few scientists would attribute it mere illusion. However, since the achievements of philosophical materialism are not yet commonly known,[150] scientists have to invent their home-made philosophies to serve as a methodological basis of scientific research. Usually, such philosophies become a kind of re-discovered metaphysical materialism (also known as physicalism, natural-scientific materialism, pragmatism, realism *etc*). That is, the only reality a scientist can see is physical nature in the broad sense, including chemistry and physiology. There is no room for specifically

---

[150] Even in the former socialist countries, dialectical materialism and the materialistic philosophy of history have never been properly taught in schools and universities, remaining a formal obligation rather than a source of practical ideas.

biological phenomena, nothing to say about the essentially social effects. In such a narrowed materialism, consciousness is deemed to be an attribute of an individual physical body, like mass, chemical composition or organic function. This poses an artificial problem of deriving conscious experience from the physical and organic organization of the individual. The problem is known as the hard problem of consciousness (David Chalmers), and it is hard indeed; moreover, it is definitely unsolvable, because subjectivity cannot be deduced from physical or biological phenomena, it requires a qualitatively different approach. No matter how completely we know the functions of the brain and human body, no matter how deep into molecular and atomic structures we delve, we won't get closer to understanding consciousness, since it is not there. Subjectivity does not require any special physics or physiology; it only organizes the ordinary physical world in a special way.

In the "physical" (or "neurological") studies of consciousness,[151] the mind is treated as an immediate consequence of the complex organization of an individual, a "higher-order" property that can be derived from the individual's construction, just like the chemical properties of molecules are attributed to their atomic composition in the traditional elementary school chemistry. However, such an approach is not generally applicable even in natural sciences, and its inadequacy is even more pronounced whenever human activity is concerned. No doubt, certain features of human behavior can, at least in principle, be satisfactorily described in this way; but these are the so called "easy" problems of consciousness, pertaining to its specific implementations rather than its understanding as the distinctive quality of the subject.

That is why some researchers losing hope to explain consciousness through the already known physical forces come to imagining new physical entities that should be responsible for the transformation of a physical body into an active personality. They speculate upon phantasmal quantum microtubules in the brain (Penrose), or invent sub-elementary relativistic particles produced from vacuum to start a new causal sequence in an act of voluntary decision (the Argentinean electrophysiological school). In some cases, such imaginary phenomena can provide fair working metaphors—however, they are far from true understanding, only rephrasing the old idea of a homunculus in one's head.

Repeated attempts to model conscious behavior reproducing its outer

---

[151] The most influential school of this type is known as "neuroscience".

manifestations constitute the other aspect of the metaphysical stand. This way will never produce artificial consciousness; it can only lead to artificial intelligence. Thus, a computer program can mimic human behavior within a specific class of situations in any minute detail, but this won't make the computer conscious, and the similarity will remain superficial. Confusing such partial models with their prototype is similar to taking an audio record for live orchestra, or a video record for live performance.[152]

To avoid spurious problems and blind search, the hierarchical organization of the world must be acknowledged, with qualitatively different levels of inanimate motion, life and conscious activity. It is no way to reduce the higher levels of hierarchy to the combinations of lower level phenomena; each level must be treated according to its nature.

Belonging to the social level of reflection, consciousness is not bound to an individual; rather, it is a collective (social) effect manifesting itself through individual activity, and organic processes accompanying it. That is, consciousness *cannot* be attributed to the internal organization of the individual, and even less to the individual's brain; the fabulous "neural correlates of consciousness" should not be considered as more than they really are: only correlation. Asking about brain structures responsible for consciousness (or its specific aspects) is no more legal than trying, say, to determine which of the two electrons in helium's $(2s2p)^1P$ state is $2s$, and which is $2p$; there is a collective effect, and any partial contributions are merged in it. Consciousness belongs to the individual no more than the form of a wave packet is contained in any of the constituent partial waves.

Specialized neural cells and structures have been extensively described in the literature, and, of course, this knowledge must be taken into account when speaking about conscious experience, perception, creativity *etc*. However, the crucial point of the hierarchical approach is that the details of neural organization are not important for consciousness, as soon as a definite complexity level is achieved. For instance, the same hierarchy of perception may be physically implemented in many ways,[153] up to quite different from terrestrial organic forms. This insures the unity of reason in the Universe. Moreover, it is exactly this independence of the organic processes that

---

[152] Many people still take novels and movies for real life; similarly, politicians are often judged by what they say rather than by what they do.

[153] Considering its empirical forms, the same perception can occur with or without the cultural context. In the former case it is called perception proper; in the latter case it is mere illusion.

characterizes the level of subjectivity as such. A brief glance to the development of humanity provides enough evidence of the growing universality of human behavior gradually overcoming biological restrictions.

The same scheme $S \to C \to R$ is applicable to all the levels of inner activity; however, it describes different phenomena on different levels. Thus, since the subject is originally a living organism, $S$ and $R$ stay for *stimulus* and *reaction*, while $C$ describes the physiological mechanism of reflex. It is here that psychological research starts, and the origin of many psychological phenomena in humans can be traced to the analogs in the animal psychology. Nevertheless, this is not yet a specifically subjective level, and one can only refer to it a low level mechanism, the basis of individuality. Though neural processes do not constitute the material basis of the subject (which only exists as a cultural phenomenon), they ensure that the subject can be *projected* onto an individual representative of the biological species *homo sapiens*, as if subjectivity could be contained in an organic body. Individuals are conscious in the sense that they commute the physical processes necessary for universal mediation, but this individual consciousness is not due to a hierarchy of reflexes, however complex, but to the modification of reflexes by the cultural environment, by the individual's inclusion in the society.

There are numerous examples of how physically normal people put outside the cultural environment and devoid of communication with other people could not develop as conscious being or lost their earlier cultural acquirements. Though the functionality of the brain remained essentially the same, organic premises were obviously not enough for conscious experience, save the most primitive intelligence. On the contrary, there is evidence that higher animals can develop quasi-conscious forms of behavior when they are included in a well-structured cultural environment.

Virtually, consciousness does not need to be embodied in any particular material form, and the biological forms we commonly observe are only one possibility along with many others. This supports the ancient dream of producing a sort of artificial subjectivity on a quite different component basis; such an accomplishment, however, would not mean more understanding, since the ability to give birth to a "thinking machine" (or a child) does not necessarily imply any knowledge of how it operates.

Of course, the components must be appropriate for the task, and their organization must allow the manifestations of consciousness. Thus, solitons cannot appear in a linear medium; also, they cannot be detected in a highly dispersive medium because they would not be stable enough. Non-linearity is to be controlled using a resonance mechanism, to select a few distinct levels (operation modes). A change in the degree of non-linearity may lead to a drastic change in the observable structures; for instance, in aesthetic perception it results in the transition from one scale to another, changing the whole perceptual hierarchy.[154]

The biological species *homo sapiens* seems to support some manifestations of reason, though one could question the sufficiency of this implementation and suggest improvements in people's life to make them more human.

Presumably, modern computers are not intelligent enough for supporting consciousness; in principle, nothing forbids further development in this direction. There is an alternative opinion that computer technologies and programming platforms have already reached the due versatility, and it is communication between computers and productive activity that are lacking. The truth of each stand is to be revealed in practical activity.

According to the hierarchical understanding of consciousness, the keyword is *universality*. While any special models of human behavior are restricted to specific classes of situations, or particular aspects of behavior, real consciousness is something that unites all those partial aspects into a whole. The integrity of the subject is an objectively formed level of the hierarchy of integrity forms pertaining to the world in general; that is, subjectivity has its roots in the other objective forms (and thus can be modeled by them), though it is qualitatively different from them (and hence cannot be reduced to them completely).

The basic mechanism of universality in human behavior is *productive activity*, with the stress on (re)producing the world rather than merely being in contact with it. In particular, the subject can (and virtually must) (re)produce itself at any level. The other side of this active universality is *sociality*: the awareness of one's self comes from the awareness of the others' products, the ability to distinguish the "artificial" from the "natural", including the observation of society-induced features in the individual.

People can be said to be conscious inasmuch they can distinguish themselves from their biological bodies, and the truly subjective existence quite often begins only after the physical death, which has been speculated upon by religion. In certain cases, a person can exist without a biological

---

[154] L. V. Avdeev and P. B. Ivanov "A Mathematical Model of Scale Perception" *Journal of Moscow Physical Society*, **3**, 331–353 (1993)

body at all; one could mention various fiction characters (like Alice or Mr. Pickwick), mythical heroes (like Jesus Christ or Hercules), group aliases (like Kozma Prutkov or Nicolas Bourbaqui), *etc*. Moreover, in their "live" form, most people are not much noticeable, they have to be extorted from the hierarchy of habitual relations for the others to appreciate their importance and uniqueness.

## *Consciousness and the brain*

Phenomena like the mind, reason, or consciousness, appear on a certain stage of development, forming a specific level of hierarchy, namely, the social level. Organic formations can only be a premise of consciousness, the way of its implementation, but not its content.[155] That is, the study of human physiology and in particular the structure and functions of the brain gives few clues to consciousness, like the design of an Intel processor says little about MS Windows, or Linux, that may run on an Intel computer.[156] Consequently, one can hardly derive consciousness from neural processes; rather, the specificity of neural processes in humans must be deduced from the cultural development and sociality. Consciousness is present in the brain (or in any cerebral subsystem) only as a way of coordinating mental processes imposed by the specific social environment, in which an individual has to live and act.

Though conscious human behavior is not a result of neural processes, it cannot, at the present stage of development, do without them. Primarily, consciousness refers to the person's co-operation and communication with the others, and hence the person's social position and the corresponding hierarchy of available activities. Neural processes necessarily accompany every activity, but they are not its source or cause. To assume the latter would be the same as to think that the plants produce oxygen just because their leaves are green, or to say that a pie is the result of a hot oven... Such associations only fix a relation between two observed phenomena, but that

---

[155] The same holds fort the levels of "physical" existence and life.

[156] This is the well know problem of reverse engineering. Sometimes, heavily using the context, it can be possible to derive the purpose of a computer program analyzing the code; more often, however, it is the programmer's comments that help. From knowledge of components, one can guess the idea of the whole; however, this possibility is entirely dependent on the commonality of cultural environment, including the standard ways of combining the standard components.

relation is due to something else, which is the origin of the both.

The relative independence of consciousness from the brain is supported by numerous facts from everyday life and scientific observations. Thus, many people have at least once experienced the feeling of losing their true self and acting like somebody else. Quite often the modern society makes its members feel themselves as if they were the wheels of a huge mechanism unable to act according to their own will and inclinations. Many people feel and behave differently in different companies or situations, up to becoming unrecognizable by their acquaintances. The altered states of consciousness and mental diseases provide examples of multiple personalities supported by the same human body, and the same brain. Finally, there are well known cases of intentional translocation from one personality to another: writers and actors may truly live by the imaginary experiences of the characters they invent. Sometimes several people act under a common alias, thus giving life to a person that had never existed in flesh and blood; however, the audience may be unaware of the artificial nature of such a group subject: thus, in Russia, everybody knows Kozma Prutkov, but few people can name the three poets that wrote for him; similarly, mathematicians are well acquainted with Nicolas Bourbaki, knowing nothing about the individuals who used that alias.

The important corollary is that subjectivity and consciousness do not require any specific organic forms for their implementation, and there may be forms of reason based on quite different physiology, and even different physical embodiment. Nevertheless, consciousness cannot exist without being implemented in complex enough biological systems, and it needs a kind of brain—or rather multiple brains joined by societal links. Probably, computer systems will gradually develop to the necessary level of complexity, to support non-human forms of consciousness; however, one cannot expect a computer behave like a conscious being until the social factors of subjectivity are reproduced.

Consciousness is no more a result of the brain's functioning than the velocity of a falling stone is related to its mineral composition. Nobody doubts that the laws of motion may depend on the properties of the moving bodies (like in the well known case of parametric resonance)—however, any motion can only be in relation to a definite reference frame, and it is external interactions of the bodies that specify both the kinematics and dynamics within the system. For conscious activity, its essence is in the social motion, while its specific forms may be due to organic influences.

Though consciousness is relatively independent of the particulars of brain physiology, or any other possible implementation, participating in a collective motion creates a special environment for an individual, directing its development to support certain physiological formations and suppress the infinity of other possibilities. Thus culture becomes projected onto the brain, regulating the relationships between its various subsystems. This accounts for the well known fact that the same behavior may be accompanied by quite different patterns of neural activity, and this diversity grows with the degree of subjectivity increasing. Subjective states are different from physiological or physical states of the organism—they require considering a wider range of events, involving the cultural environment of the individual.

### *Localization of mental processes*

While consciousness cannot be found in a single organism, and in particular anywhere in the brain, some mental processes can be localized in the cultural space, with the subject occupying a definite place in this space (figure 5). One can consider the subject's dynamics in the cultural space, and describe social influences as external forces.

Considering neural processes in humans as a component of a particular implementation of subjectivity, one could investigate the possibility of correspondence between cerebral structures and the structure of activity. An adequate description of psychic processes will avoid direct association of a specific psychic phenomenon with a single neural pattern, or an area in the brain, since many neural patterns may implement the same inner activity, and the same neural processes may serve for quite different mental acts. It is in the global arrangement of physiological processes in the brain (and the organism in general) that consciousness can manifest itself.

Like originally separate organisms developing in permanent contact adapt to each other and become the organs of a higher-level organism, human bodies involved in conscious activity adapt to the cultural environment that forms the non-organic body of the subject. In human phylogeny and ontogenesis, only a few of the possible modes of brain functioning can be selected, and the organization of inner processes is dictated by the needs of external activity. The evolution of the human brain had to retain some traces of this correlation in the overall structure of the brain, though it could not produce a rigid structure since the very idea of subjectivity implies diversity and overcoming physiological constraints.

In general, the structure of the brain as it has formed in humans reveals three major asymmetries: occipital-frontal, left-right, subcortical-cortical. These are the space-like dimensions; also, there is a time-like dimension related to the development of the brain with the formation of primary, secondary and tertiary areas clearly distinct in all the spatial dimensions of the brain.[157] It should be noted that the asymmetry of the brain has been becoming more pronounced in the process of biological development, from the lowest to higher animals; the biosocial development of primitive humans significantly contributed to this differentiation, which is related to structural differentiation of activity. Thus, syncretic reactivity expressed by the scheme $O \to (SR) \to O'$ (with $S$ and $R$ being the same organic structure taken either as passive, experiencing the influence from object $O$, or active, influencing object $O'$) does not imply much cerebral differentiation (and even the presence of the neural system at all), while the complete sensorimotor scheme $O \to (S \to C \to R) \to O'$ says that there is a special structure S for receiving external influences, and a special structure $R$ for effectuating the outer behavior, with some inner structure $C$ to transform stimuli into reactions; evolution from simple forms of reactivity to complex reflexes is hence accompanied with structural (and functional) separation of sensory and motor zones, and formation of specialized mediating structures. The same process has lead to the separation of the two principal "logical" functions of the brain: folding the multimodal image of the world into an internal integrity $(S \to C)$, and the unfolding of a syncretic internal core into a discrete hierarchical structure $(C \to R)$; this corresponds to the lateralization of neural activity with cyclic translation of the same pattern from the left to right hemisphere, and back, within a single mental act.

The cerebral representation of the schemes of inner activity is different in different "time points," that is, in the structures of different historical age. Thus, the low-level sensorimotor scheme $S \to C \to R$ is cerebrally implemented as direct links from sensory to motor areas mediated by the special sensorimotor zones physically located between the sensory and motor zones. On a higher level, there is yet another way of mediation, through the lobe zones, and the spatial order in the occipital-frontal dimension becomes inverted. From the lower-level view, this looks like the formation of the indirect links of the $S \Rightarrow R$ type. Space and time are strongly coupled in the brain, and the geometry of the cerebral space-time is essentially non-

---

[157] A. R. Luria *Foundations of neuropsychology* (Moscow State University Press, 1973)

Euclidean. However, this is essentially a 3+1 geometry locally similar to that of the physical space-time, though in a quite different space, namely, that of neural dynamics.

The spatial and temporal characteristics of the cerebral mechanisms of mental processes are not necessarily related to the space- or time-like parameters of the mental processes themselves, since the latter unfold in a different space and cannot be localized in neither physical nor physiological sense. Any association of a mental phenomenon with a brain structure can only be treated as a simplification valid for the conditions in which one mechanism dominates over the others, that is, for a single unfolding of the hierarchy. Thus, one cannot say that certain kinds of mentality (and corresponding types of behavior) are related to definite areas in the brain—rather, every type of behavior implies the participation of every part of the brain, and it is the mode of their interaction that matters. As a consequence of a physical or social trauma, this integrity of neural activity can be broken, with some regions of the brain dynamically or physically isolated from the others; this manifests itself as a mental disease.

Generally, higher-level processes are neurally represented through coordinated operation of various low-level mechanisms. For instance, perception involves the operations of categorization, structuring the sensations and bringing them under the pre-established patterns of activity; this means that the activation of the sensory and motor zones in the brain has to be synchronized through specific frontal areas, depending on the level of perception (and the current sets). The dominance of, say, back-right cortex will make perception impossible, reducing it to unstructured sensations; on the contrary, the dominance of a motor zone means disintegration of perception because of a poor correspondence of the inner representations to the external situation.

Normal functioning of the brain within some inner or outer activity implies absence of a fixed dominant region, but rather a kind of rotation, with the different parts of the brain dominating in an activity-determined sequence. When the activity becomes too rigid, only a few of the possible trajectory classes remain accessible, which results in more frequently switching between them and "jerky" behavior indicating a problem situation. Generally, there are individual styles of behavior implying certain constraints on psychodynamics and neurodynamics, though still allowing much mental flexibility; every such style can be associated with a class of cerebral processes, that is, a number of typical sequences in the interaction of different parts of the brain.[158] It is well known that too much adherence to a single behavioral style can undermine the efficiency of activity and lead to conflicts. In the extreme, this behavioral confinement may develop into neurosis, which is accompanied by stagnant neural dynamics, the dominance of one of the possible phase trajectories. When there are many trajectories that connect the starting point with the goal, impossibility of actualization of one of them does not hinder the efficiency of action; if there is only one possible path, any obstacle (*e.g.* a trauma) will be able to break the chain of cerebral commutations and form isolated excitation zones in the brain, with complete destruction of consciousness and subjectivity.

## Space and Time

From the viewpoint of the hierarchical approach, time is related to hierarchical development, the "vertical" growth of hierarchy. Interaction with other hierarchies and their mutual reflection is the basic mechanism of this growth. Hierarchical reflection is also a hierarchy, and it can be unfolded in a sequence of individual acts of reflection, constituting the hierarchy's reproduction cycle. Time can be quantitatively measured by the number of full cycles. Lower level motion in the same hierarchy is characterized by its own period of reproduction, and hence a different time scale.

The structural aspects of a thing provide its "static" picture, while its systemic aspects bring in the idea of "dynamics". Systemic dynamics is an inverse of the system's structure, and systemic description is complementary to structural description. This leads to the relativity of the distinction, so that structural aspects may become functional in a different context, and conversely, systemic features can be treated in a structural way. Such transformations are well known in physics, where time coordinate is like spatial coordinates in any respect.

However, time coordinate does not fully represent time, and it is only in the hierarchical approach that historical time can be understood as different from mere systemic dynamics. Developmental study synthesizes both static and dynamic views, regarding a thing that changes, while being the very same thing.

---

[158] A style of behavior could be defined as a class of the possible trajectories of the top of the hierarchy in the process of refolding.

Though space and time are the general forms of any motion, they differently appear on different levels of the hierarchy of reflection. Thus, physical time is syncretic, and, in many cases, it can be well represented by a number. In more complex systems, internal modes of motion must be accounted for, so that there are a number of interfering processes occurring in different time scales. However, in the inanimate world, all such time scales are equivalent, and there is no preferred direction of development. In other words, physical time is reversible due to its essential locality. Various theories of irreversible physical processes had to somehow (often implicitly) introduce non-locality to violate the reversibility of time.

On the level of life, the distinction between the organism and its environment becomes more pronounced due to that every organism is a lifted-up result of biological evolution and a representative of a species. In this latter quality, the organism is no longer local, and the direction of development is closely related to the type of the organism's relations with environment, which becomes reflected in the structure of the organism through biological evolution. The "spatial" distinction of the external and the internal thus becomes closely related to the distinction of the future and the past. In animals, this distinction becomes quite prominent in behavior, being the core of a *reflex*, a natural behavioral junction between past experience (either genetic or individual) and the development of the current situation into the future. However, an animal normally does not reflect its own reflexes, so that the past is immediately juxtaposed in the animal's behavior with the future, without mediation; the present is a single moment for the animal. It is only with the first glimpses of consciousness that the syncretic connection of the past and the future gets broken, and the present comes to existence as a kind of structured mediation.

On the level of activity, internal reproduction of the world becomes universal; every outer act is accompanied by hierarchically organized inner activity, and every internal act finds reflection in public behavior. The natural measure of time is obviously related to the cycle of objective and subjective reproduction, activity.

The levels of operation and activity are connected through the mediating level of action. This hierarchy is also applicable temporal relations in conscious behavior. Thus, the present is correlated with the focus of awareness, the direction of one's attention *now and here*. This is the definition of the goal of an action. The past is reflected in the subject as folded (interiorized) experience, which formerly was in consciousness but is no longer present, being merged with the subconscious, on the operation level. Recollection is hence readily interpreted as unfolding the hierarchy. The future is *yet* absent, but it is present as what *will* be, and hence belongs to the *zone of imminent development*, as defined by the current activity.

Unlike in animals, people's inner life is universally reflective, and it is this reflectivity that makes the present last, unfolding it in an internal process between the conscious impression and conscious response. This internal process duplicates the outer life as an act of subjectively living through the situation, *experiencing* it. Consequently, experiences in the proper sense can exist only in conscious beings; for inanimate things or animals, this word is used metaphorically.[159] Things and animals are subject to external influences, and they interact with their environment, but they do not *participate* in an external process like humans do.

The way of *experiencing* time can be different from its developmental relatedness to the levels of activity, action and operation. Thus, psychological orientation is more likely to be directed to the higher level rather than kept within the same level: for instance, an operation is associated with an intention, which relates it to the present rather than to the past; similarly, an action is psychologically directed to the future, while an activity requires an orientation to the folded activity of a higher level (social experience), often syncretically represented by various regulations, norms, rules *etc* referring to the past.

Due to relative independence of the focus of awareness from the current activity, it can shift from one level to another, producing different subjective representations of the same conscious act. For instance, the activity of *listening* folds into the action of *attending* focusing awareness on certain aspects of the whole sound pattern; in the most folded form we deal with the operation of *hearing*. However, listening does not mean *hearing* unless there is an operation level within each act of attending making it structured enough to become a conscious experience. While you simply listen, you don't hear—this is a well known paradox.

Considering inner activity, we come to the notion of subjective space and time different from physical space-time, as well as from biological space and

---

[159] The same holds for the common words like "behave", "act", "perceive" *etc*. In a strict language, one would use different terms for similar phenomena in the inanimate world, life and subjective mediation. However, any object implies the presence of the subject, and hence natural language is essentially metaphorical; rigor and objectivity can only maintained locally, in the same position of hierarchy.

time. Subjective space and time can have topology different from spatial and temporal structures on other levels. Everybody knows how a distant event can be closer that something happening right here, and memories can be stronger than current perceptions. Since the subject reflects the world in an objective way, the internal model of physical and biological space-time will correlate with the lower levels; however, this model does not coincide with subjective space and time, which can also be subjectively reflected.

## Development of Consciousness

Considering the origin of consciousness, one has to investigate how the phenomenon of consciousness is related to the physiology and other lower-level phenomena, as well as study the history of consciousness, indicating the necessary premises and the principal paths of development.

Every subject is primarily an object, but the very definition of subjectivity as universal mediation characterizes it as an objective property external to the object, so that no body, however complex, can "contain" subjectivity. Material implementations are necessary for subjectivity, but they will always remain relative and incomplete. The subject cannot be reduced to a single body or a system of bodies, however complex. Still, some level of complexity is necessary for material motion to support life, and only the highest forms of life can support consciousness.

Reflection is the universal mechanism of development; on the level of life it takes the form of external reflection, mutual influence of the organism and its environment. Consequently, both the organism must be flexible enough to develop diverse reactions to external stimuli, and the environment must be complex enough to allow flexible co-operation of different organisms. However, the level of cooperation required for consciousness differs from mere symbiosis and ecosystem formation common in the biosphere. In addition to external reflection, there is internal reflection, the subject playing the role of environment for itself. This can only be achieved through directed rearrangement of the inanimate and living world, its transformation into the hierarchy of products, culture. Reproducing the world in cultural phenomena, the subject controls its own development, interiorizing the products of social activity, which thus become "immersed" into the subject, making external phenomena a part of internal life. Such "representedness" of the culture in the subject requires lifted-up retention of the cultural phenomena both directly in the immediate environment of the individual and in a mediated form, as influences on inner dynamics (reminiscences). Culturally imprinted, these traces become a part of the extended body of the subject, which is hierarchically organized in accordance with the organization of outer activity,[160] virtually determining the subject's singularity.

The higher levels of hierarchy always develop on the basis of the lower levels; that is, the common physical world is enough to produce life, and consciousness does not imply any "subtle bodies" different from physical, chemical or biological bodies. It is only a specific arrangement of material bodies that makes them function as an organism, or a subject.

The system of material bodies that can support consciousness contains the organic bodies of the representatives of the genus *homo sapiens* as a part of a wider material system, the "non-organic body",[161] including both inanimate things and living organisms; however, such outer components play the role of the non-organic body of the subject inasmuch they are used as its outer organs (tools, instruments), otherwise belonging to their appropriate levels of reflection. In particular, humans can belong to the non-organic bodies of other subjects, thus losing their specificity of the carriers of consciousness.

Since the primitive historical forms of subjectivity have been left behind, the non-organic body is dominating over the biological body in shaping the subject's behavior and thoughts. It is from that non-organic body that the absolute majority of the people's motives originate, and it is those outer organs that are mostly used by the people to interact with the world. The non-organic bodies tend to expand with cultural development, and the disproportion between the organic and non-organic component is bound to increase. The development of thus embodied subjects is in no way limited by biological evolution and genetic laws,[162] and it will finally be completely controlled by the conscious will, since the universality of subjective mediation implies the mediation of one's own embodiments too.

The interactions of the non-organic body are as important for a person as

---

[160] It is important that consciousness develops together with its field of activity, the subject develops together with the object.

[161] "Non-organic" means "not of the organism, biological body"—it does not imply that all the parts of the non-organic body must be inorganic, though some technological trends might suggest this idea.

[162] Independence of the organic body can serve as a measure of consciousness development in humans.

directly acting on his or her organic body, and they are felt by the person as keenly. That is why traditional psychotherapy mainly based on organic reactions and local influences always risks to be completely neutralized, or even turned into its opposite, by the changes (or absence of changes) in the circumstances of the patient's life. The larger the body, the greater the period of reflection,[163] and the difference in reflection rates may allow a short-term therapeutic influence on the organic body of a person to lead to certain psychological changes—however, these traces won't last for long without social support.

Since development of the subject is mainly determined by its non-organic body, physiological peculiarities are not very important, provided they are not accentuated by the society (and thus included in the non-organic body). A certain level of complexity achieved by the human organism is enough to allow its usage as one of the organs (instruments, tools) of the subject, but possessing a perfect organism does not guarantee well-developed subjectivity. With properly organized education, children born blind, deaf and numb can become normal members of the society, while organically normal children that grew among animals fail to develop into the conscious beings.

In hierarchical development, lower-level features change whenever a new higher-level feature forms; immersed in a different environment, they have to adapt to new conditions, which demands functional and structures modifications. For humans, cultural development plays the role of selection factor for organic development; changes in the way life induced by the culture will definitely influence metabolism and shift biological preferences. The structure and functions of the brain are especially sensitive to the new patterns of activity appearing in economic and social development; however, such organic changes happen in the biological scale—that is, very slowly compared to cultural changes. Human physiology remains almost the same through centuries, while human ways of life (and hence human minds) become drastically different.[164]

Today, we have a good model of such a development process. Computer hardware and software develop in a pace that allows observation of several

---

[163] F. J. Dyson "Time without end: Physics and biology in an open universe" *Reviews of Modern Physics*, **51**, 447–460 (1979)

[164] Humans have already come to the level of technological development that allows direct modification of the human organism according to cultural needs. Birth and death are to be put under conscious control, and the traditionally biological modes of reproduction have to give way to an industrial process.

revolutionary changes during a single lifetime. We can see how the same hardware (the "organic" part) is used to implement increasingly complex functions, and how hardware changes to yield to the pressure of expanding needs, so that a part of earlier software-emulated functions becomes hard-encoded, with increase in efficiency. This opens new directions for software development, and so on.

Two opposite principles determine the development of the humanity as a collective subject. On one hand, every trace of subjectivity, once historically formed, must stay forever due to subjective universality; on the other hand, nothing can be eternal in history, and every historical form has to die when it is no longer supported by objective necessity. The history of consciousness must hence be the unity of stability and mutability, with the universal core undergoing incessant transformations.

Historically, the first glimpses of conscious behavior were interwoven with syncretic productivity of primitive humans; there could be no special reflective activities characteristic of the later stages of cultural development. The subjective character of the early forms of human activity could only manifest itself in the new types of inner reflection representing the abstract universality of arising consciousness; though such inner activity must reflect some aspects of observable behavior, these objective prototypes are originally a part of apparently animal way of life, and any specifically human qualities initially develop in animal-like forms. This is often valid for modern humans as well, since every person is engaged in a hierarchy of activities including those aimed to supporting life and survival; sometimes, it can be difficult to discover subjective core (inner activities) in apparently animal acts (anger, panic, conformism *etc*).

The fundamental structure of conscious activity, $\boldsymbol{O} \to \boldsymbol{S} \to \boldsymbol{P}$, is yet syncretically present in the outer behavior of early humans; it is as syncretically reflected in the triad of basic inner activities:

*contemplation* $\to$ *imitation* $\to$ *imagination*.

These activities first appear as abstract abilities, as inner premises of consciousness, serving as a bridge from animal to properly human forms of behavior.

Contemplation is the most primitive form of internal activity comparable to the spontaneous reactivity of an amoeba. In contrast to mere postponed reaction, contemplation is not centered on the object, being too diffused and generalized. In contemplation, all the external stimuli are brought to the same internal pattern, broadly identified. Contemplation is the germ of abstract thought in a primitive human, or a higher animal[165]; allowing to separate internal activity from external, making them relatively independent.

Imitation as internal activity directly corresponds to outer imitation, which is the ancient mechanism of socialization, the earliest form of collective operation. Compared to contemplation, it represents a higher degree of abstraction, since it does not require direct stimulation. By its form, imitation seems to refer to some real act; however, this concreteness of imitation as internal activity is utterly superficial, since the content of imitation is much more abstract than diffused reflexes of contemplation. Internal imitation can rarely be observed in animals, despite the well-known fact that external imitation is the basis of learning on the animal level.[166]

Imagination brings this hierarchy of inner abstraction to its culmination, removing all relation to experience inherent in imitation. The social component of imagination seems to be much weaker, though it is the diversity of communication with the likes that enables one to imagine anything at all. The products of imagination do not need to be purely mental: here, the "internal" character of activity only means that it mediates a mental act, which can occur on either individual or group level.

Dreams could be considered as an important physiological premise of imagination. Though imagination is often treated as a kind of "daydream", this is no more than a metaphor; in contrast to the "stochastic" imaging of dreams, imagination is well organized, producing forms that are determined by the structure of activity.[167]

## Education

The self-reproduction of the subject on the global level is complemented

---

[165] Some animals can exhibit quite definite signs of contemplation, especially when brought up in a human environment. Cats provide one of the most pure cases. Much of the cats' quiet musing about the happenings around seems to have been borrowed by some oriental meditation techniques.

[166] For early human, art was mainly a material mediation of internal imitation. This stresses the difference between external and internal imitation: the former is a functionally exact reproduction, while the latter requires only *subjective* similarity.

[167] One could characterize imagination as "controlled dream", and there are inner states intermediate between dreams and imagination. This, lies in the basis of the psychological techniques that employ gradually gaining control over one's dreams as a means of self-regulation and stimulating creativity.

by the activities directed to imposing subjectivity on the individual members of the society; education is a necessary component of the subject's universality, which implies conscious construction of conscious beings.

Education also complements the physical reproduction of individuals, which is also gradually coming under conscious control. The important feature of education is that its basic principles and forms do not significantly depend on the bodily implementation of the subject, being primarily determined by the cultural formation. Human physiology will change, a kind of artificial intelligence will be created sometime; this does not influence the necessity of education to transform the raw organic or pseudo-organic material into the agent of conscious activity. This process is essentially social.

One cannot influence consciousness otherwise than through the body, since no consciousness can exist without material support. However, one cannot influence consciousness only through the body, since consciousness does not exist within the body, being relatively independent of a particular implementation. Similarly, consciousness cannot express itself otherwise than through material acts, and it can never be reduced to the bodily process. The solution of this contradiction must be sought for in the mechanisms of interiorization/exteriorization, when the relations and interactions between the objects become the objects of a special kind constituting the "interior" of the subject.

Education is not mere conditioning, training, developing social habits. The mechanism of conditional reflex is entirely animal; it is contrary to free will that is required for conscious action.

A newly born child is not yet a conscious being; it does not even have the necessary inner activities to support consciousness. Traditional dualistic view of the mind and its content as separate entities presents education as merely filling the ready-made mind with some culturally determined contents. In reality, the content of the mind depends on the level of mind's development, and socialization is not passive, it demands social behavior of the pupil in addition to external guidance. Of course, a number of operations must be simply learnt, to "tune" the organism to cultural modes of functioning. However, this is only a preliminary stage of socialization, which cannot produce consciousness as such.

The general law of learning is conversion of hierarchy. First, the most general command is formed, which cannot be fulfilled. Then it unfolds into a number of partial activities, and thus repeatedly until the level of detail is enough to act. This is the stage of unfolding. After the simple actions have been trained enough, they fold into operations, and higher-level actions are being learned. Thus the hierarchy folds into a simpler structure. However, it can fold differently, and the result may be far from the original idea.

Individual development, as usual, mimics phylogenic development. There is no "final" state in consciousness formation, and people can only be considered as conscious judging by the level of presence of the reasonable component in their behavior.

Stages in individual development do not necessarily coincide with any organic development stages. Originally, for early humans, there is high correlation between the two, but, with time, it becomes less pronounced in human history. Even when a psychological crisis accompanies a biological shift, psychological changes do not depend on mere physiology, they are always socially driven.

Basically, education can proceed on the syncretic, analytical or synthetic levels. Syncretic education is important as an element of cultural experience; it is directed to reproduction of the standard ways of action and thought, imposing them on the person since early childhood. Analytical education is a special activity, in which teachers, professors *etc* explicitly or implicitly (by example) transfer their cultural baggage to pupils, students *etc*. Quite often, this activity assumes institutionalized forms (schools, institutions *etc*). In synthetic education, like on the syncretic level, cultural interaction within a common activity is the main mechanism of scheme transfer; however, this transfer is intended by the very organization of the common activity (work).

The forms of educations depend on the cultural formation. The general direction of their development coincides with the overall direction of the development of economy: from individual to social activity. Primitive education is possible in a family; however, it is only the society as a whole that can suggest an environment rich enough for subjective universality. Gradually, with reorganization of the society towards more cultural determination, hierarchical planning and regulation, the very distinction of separate social groups will go to the past, and education will become entirely social. This is the only way to adequately account for individual capabilities and predispositions; in a class society, individual approach in education is limited by economic restrictions, and most talents get lost because of impenetrable social barriers.

Considering individual development as socially determined construction of the person's hierarchy (including physical, psychological and social respects), one could question the overall level of integrity thus produced. The internal contradictions on the present cultural formation often become reflected in various personal problems; the two aspects of a well-balanced education are health and maturity. Poor living conditions of the vast majority of the population lead to both physical and mental misbalance, and virtually disease. On the other hand, the poor suffer from the limited access to cultural heritage, which results in deficient subjectivity, lack of motivation and responsibility. Lower classes do not get proper education to grow into really conscious members of society; they remain under external control, mentally sticking in the childhood.

Different aspects of education are not rigidly correlated. One can be physically healthy, but mentally deficient, with socially distorted personality. Conversely, an invalid can be a healthy and mature personality. There are both physically and mentally healthy people, who do not possess any higher interests, just living with what they have, collecting goods and pleasures. Such people can hardly be said to possess consciousness and reason.

Compliance with the definition of the subject, universal mediation, is the main criterion of personal adequacy. One represents universal subjectivity acting in a universal way, transforming nature to produce culture. Mere existence, or animal life, is not enough to become truly human.

Maturity characterizes a different aspect of personal development. In a sense, maturity is opposite to reason, since it apparently means a state of completed development, while the very definition of subjectivity is related to its extended reproduction, development. However, in the hierarchical understanding, there is no subjectivity without maturity, and maturity can admit ability of subjective development. Maturity could be characterized as a specific position of the person's hierarchy, one of possible positions, a well-developed hierarchical structure. This structure reproduces the structure of the society, and culture in general, within the present cultural formation. That is why, while subjectivity does not depend on the culture, maturity can be differently defined in different cultures.

## Consciousness and Communication

Activity as the cycle of subject-object reproduction
$$... S \to O \to S' \to O' \to S'' ...$$

is the unity of the object (productive) cycle
$$... \to O \to S \to O' \to ...$$
and the complementary subject cycle:
$$... \to S \to O \to S' \to ...$$
While the object cycle is the starting point for unfolding the inner hierarchy of the subject, the secondary reproduction cycle can be used in discussing communication between different subjects, as well as its special case, one's communication with oneself.

Basically, a communication act is exchange of objects between the subjects involved; for two subjects $S$ and $S'$, the subject cycle becomes the unity of two transactions:
$$S \to O \to S'$$
and
$$S' \to O' \to S$$

The majority of human communication is implicit, through exchange of the products of people's common activities. People consume material things to produce other material things, and the very process of consumption is already producing something, as well as any production has to consume goods. This mutual penetration of production and consumption in the cycle of cultural reproduction is the fundamental mechanism of consciousness formation, and it has two complementary aspects. First of all, material production forms a specific environment for the contacts of people with each other. People become interdependent in an economically mediated way; such relations make their behavior quite different from merely reacting to external or internal stimuli. Conscious behavior is a part of an objective process, and is organized by the regularity of this process, which is reflected in each individual as purpose and goal. Becoming a part of the society, an individual acquires additional qualities, which characterize that person as a specific active agent, a personality.

There is another principal aspect of the reproduction process. Common activity implies synchronization of individual actions, and hence a number of component actions (communication acts) serving such a synchronization. The products of these actions are called signals. Each activity develops its own hierarchy of signals, and the universality of the subject means that all kinds of physical signals can be used, transformed to carry an essentially non-physical message. Since production and consumption are the sides of the same process of cultural reproduction, every act of communication becomes exchange of signals rather than one-directional information transfer. This feature is reproduced in various artificial systems (*e.g.* computer protocols that allow many computers participate in the solution of a common task).

Obviously, conscious communication must differ from non-conscious signal exchange in the same respect in which subject mediation differs from any other mediation in the inanimate or live nature; that is, it is distinguished by its universality. The same signals as used by animals become different with humans, when they act as conscious beings. On the level of subjectivity, each signal is hierarchical, combining its immediate function of the carrier of situation-specific information with the cultural function of a materialized social relation.

Though any product can serve (and serves) for communication, not all products can do it in a universal way appropriate for specifically human communication; with time, a special product has evolved, which happens to be universal enough to synchronize any activities, being independent of them. This product, language, serves as a universal objective mediator of communication joining the subjects into a higher-level integrity in an explicit way. Unlike the other products that do not function as products outside the process of their production and consumption, language retains the traces of subjectivity in a much wider range of situations, being relatively insensitive to any particular case of communication.

The universality of consciousness allows transfer of activity schemes from one activity to another; that is, a product of one activity can represent a quite different activity within a particular cultural context, thus becoming a sign. Though almost any product can become a sign, some products (words; later symbolic expressions) are more suited for that purpose; language has developed as a universal system of such special signs. Since this system allows explicating any relations within the subject, it becomes its exteriorized form, an explicit existence of consciousness.

Development of language is another side of economic development, and consciousness, as soon as it passes beyond the most primitive forms, cannot develop without language. The universal forms that are provided by language become an instrument of self-construction and self-reconstruction for consciousness, and this circumstance may lead primitive minds to the illusion that the whole of the consciousness arises from language, and verbal activity is primary to any other activity in the culture. However, the objective necessity of language formation does not imply that consciousness cannot develop in other forms, and, in any culture, there are numerous language-like activities that may occupy a significant portion of cultural space. All such modes of communication soon become language-saturated; conversely, they influence the development of the language.

Since language is primarily related to activity exchange, it is natural to compare it to yet another well-known example of exchange, trade. As Karl Marx has demonstrated, in economy based on private appropriation, exchange of goods objectively leads to establishing a social relation between products known as *value*, and money becomes the universal mediator representing this social relation. Similar processes occur in the sphere of communication, where the role of language in human communication can be regarded much like the role of money in economy. Exactly like material production gets eventually subordinated to rotation of capital, language becomes apparently independent of the relations of material production and consumption and begins to develop on its own basis. Like a banknote is rather a representative of a social relation than mere scrap of paper, words do not mean anything on themselves, outside communication context.

The simplest scheme of subject substitution in an activity is

$$O \to S \Rightarrow S' \to P,$$

which shows how one subject takes over the role of another in producing the product $P$ from some object $O$. However, the two subjects can only act as such while mediating relations between objects, and hence this scheme implies that subject $S$ has to pass some object to subject $S'$:

$$O \to S \to C \to S' \to P.$$

The scheme $S \to C \to S'$ represents a simple communication act, provided all its components are considered as universal. The origin of language as the universal mediator of communication is, therefore, to be sought for in the development of the mechanisms of activity transfer, from the occasional to universal forms.

The first primitive form of activity transfer is not separated from activity itself. Being essentially collective, any activity implies involvement of many people (as the members of society), with their roles defined by the physical and social organization of that activity. If one of the positions becomes void, some individual will occupy it, so that the integrity of activity would be preserved. This is a semi-animal form of cooperation; for instance, wolves can take over the roles one from another while pursuing the prey, depending on their hierarchical positions in the pack. In this syncretic communication the objective need serves a signal for the other person to continue the activity along the socially standard lines. One could call it *role inheritance*.

In syncretic communication, all the individual subjects are equivalent, since any one of them can occupy the position of another. However, in complex activities, the roles of the participants become traditionally fixed, with social segregation leading to a hierarchy of communication formats.

In a simple act of activity exchange, two subjects are substituted for each other: $S \leftrightarrow S'$. This means that one person can do as well as another, in the context of the specific activity, and individuals do not differ from each other in respect to that activity. As any relation like that, this substitution implies a lifted up cycle of mediation:

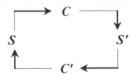

That is, to pass an activity from one person to another, $S$ needs to pass a number of material things to $S'$, as well as $S'$ needs to pass some other material things to $S$; both directions are necessary for communication. In this cycle, two complementary messages $C$ and $C'$ are equated to each other, and thus identified: $C = C'$. This identification is the primary abstraction that further leads to formation of notions and categories.

The society as a collective subject, uniting the individual subjects and groups) is impossible without a common system of communication, and a common language. Language barriers restrict people's universality, and hence they are incompatible with the development of consciousness. Consequently, the humanity will have to establish a common language to become true unity, and if any non-human forms of reason come to existence, they will have to find a common language with humans.

Since communication is the other side of any activity, and activity itself, it will necessarily exhibit hierarchical features common for all kinds of activity. Communication is commonly said to transfer information from one partner to another. However, people differ in their opinions on what kind of information is really meant. From the activity viewpoint, every "message" is a unity of discreteness and continuity, its discrete side being related to the operation level and its continuity linking it to the embedding activity. The content of the "message" (that could be called information) will always be a kind of interval, a continuous process within discrete limits, corresponding to the level of conscious action. This definition accounts for both qualitative and quantitative aspects of information. Obviously, the quantity of information depends on the level of hierarchy, and the same message will carry different quantities of information participating in different activities.

In contrast to mere interaction, or organic processes, communication is based on universal operations (potentially) available to every member of the society. The existence of such operations is the basis of the "interpretability" of the message, projecting it into the individual experience of every person. For those who do not share the same operational background the message will remain meaningless until the cultural schemes implied get somehow learned. Thus, meaningfulness can be achieved through using a definite hierarchy of scales, providing the basis for common interpretations within a specific culture.

To establish cultural links, a message must employ traditional elements, albeit unexpectedly organized. The new has to grow from within the old. Luckily, few people can invent anything absolutely uncommon, and the major problems with the meaning of the message arise from cultural incompatibility of different societies or social layers.

To efficiently communicate, different people must be involved in at least one common activity, so that the message produced by one person could have sense for the others. In other words, there must always be a context, "synchronising" the activities of the people producing the message and receiving it.[168] In this respect, information exchange is like quantum physics, which applies to essentially correlated systems observed from an upper level of dynamic hierarchy.

When two subjects communicate within some activity, they form a higher-level subject, thus becoming its parts—or, rather, participants. It is this collective subject that is the primary agent of the activity, with the inner hierarchy containing the level of individual participants. The collective subject acts in a way that is relatively independent of the individual activities of the people involved. The universality of language makes it the principal mechanism of forming such group subjects.

Though communicating people have to lose a part of their individuality, transferring it to the group, the positive effect of verbal communication is in

---

[168] For example, for a musical piece to be comprehensible to the listener, the listener must share the same scale hierarchies with the composer. Pieces written in an uncommon scale will most probably be perceived as lacking order and harmony.

the enhanced efficiency of the participant individuals,[169] increasing their universality.

Language is the universal inner mediation of the subject; in the human culture it plays a very important role. However, the communicative aspects of human activity cannot be reduced to the verbal sphere, as the very universality of the subject implies. There is no consciousness without language, and any kind of conscious behavior must be represented in a kind language. However, verbal mediation is *secondary*, being a superstructure of the object-mediated links in a joint activity. To assimilate more of the human culture one does not need language,[170] and the first steps of a child's socialization precede the first signs of verbal behavior. Verbal structures originate from common activities, which come before words as their objectively necessary predecessor. In general, the communicative system includes verbal level along with the sub-verbal and super-verbal behavior, in correspondence with the universal organization of activity.

The history of communication is the balance of two opposite processes, verbalization and "deverbalization". First, syncretic activity exchange gives way to verbal mediation; then verbal contacts become folded in non-verbal transactions. In this folded form, primary object mediation becomes transformed into pseudo-object (symbolic) mediation, mapping the language onto a similarly structured subset of the people's cultural environment. Deverbalization is related to ideation actively redesigning the world; it implies special products to mediate non-verbal communication in a universal manner, and, virtually, a whole level of culture-mediated communication analogous to super-conscious level in the hierarchy of activity.

The formation of super-verbal communication makes communication much more economical, since a long verbal message can be "compressed" using commonly accepted symbolic systems. In the simplest case, speech becomes simply encoded in a sequence of symbolic transactions. Thus oral language becomes written language, up to reducing the alphabet to a pair of distinct characters (say, "0" and "1"); any text would then be replaced by a conventional sequence of 0's and 1's. In continuation of this process, verbal communication can be folded in ideomotoric sequences, which still are able

---

[169] An analogous phenomenon can be observed on the biological level, when the voices of the animals of a certain genus form a kind of sonic background, which acts as an indicator of the stability of the environment and the absence of danger.

[170] For instance, one does not need a word for an axe to learn its usage. Mere observation and imitation is enough.

to deliver the message to the partners as reliably as (or even more reliably than) explicit verbalization.

## Altered States of Consciousness

In any hierarchy, there is no abyss between the levels, and every two substructures get connected through a number of intermediate formations. In particular, the boundary between consciousness and the unconscious implies a number of boundary states, which resemble conscious actions in that the person is, to some extent, capable of apparently intentional operation, but which are much closer to the unconscious by their content. Such positions of the hierarchy of activity are known as *altered states of consciousness* (ASC). Among the common ASCs, one could list lucid dreams, narcotic intoxication, near-death experiences, various kinds of trance, peripheral comas *etc*. Using special meditation techniques people can be trained to enter some types of ASC quickly and in a reproducible manner, retaining the ability to report their experiences while being in the ASC. However, many ASC occur under unusual conditions, significantly limiting operation with external objects and communication; all information about subjective experiences can only come from the post-ASC reports, which may significantly distort the picture (*e.g.* for lack of the adequate means of expression). The feasibility of scientific study of ASC may hence seem doubtful, or at least very limited.

However, ASC are of great interest for consciousness studies because they provide an insight into the mechanisms of hierarchical conversion in activity, transformation of a person's conscious experiences into unconscious traces and tendencies. To study such transitory states, one can resort to the standard "scattering" techniques widely used in natural sciences: comparing the products left by a person in an ASC in the external world with those produced in a similar activity that did not involve ASC, one can discover qualitative changes of the resonance nature allowing to make assumptions about the influence of the ASC on the person's internal activity. A specific modification of this approach could be used to discover the traces of ASC in the conscious agent, on the level of individuality, personality or sociality. One can admit that those who experienced an ASC will exhibit slightly different performance in the same socially imposed role; that is, engaging the ASC carrier in a relatively neutral activity controlled by the experimenter, we could detect the shifts in behavior caused by recent experiences and thus construct a kind of a scale to measure the effect of the ASC.[171]

Convertibility of hierarchies provides yet another way of studying ASC, which is based on the relativity of distinction of elements and links in a hierarchical structure. That is, for each activity, with a definite separation of the conscious and unconscious levels, there is a dual activity, which the vertical links corresponding to the levels of the initial activity, and the levels corresponding to former links. The processes of rationalization and motivation could serve as examples of activities originating from hierarchical links. Studying such exteriorizations of inner motion, one will discover numerous types of mediation related to ASC.

There are ASC that can be discovered in any normal behavior. Since the focus of awareness is relatively independent of the current activity, it can drift from one level to another, producing different subjective representations of the same conscious act. For instance, in sound perception, the level of conscious action is related to the act of *attending*; however, one might as well focus on *listening* (activity) or *hearing* (operation), thus shifting consciousness to an altered state, which is not stable and must relax to a regular structure via hierarchical conversion. The period of such relaxation may be measurable; in the meantime between the restoration of the focus of awareness, the person experiences an ASC subjectively felt as distraction, or "rugged" perception.

Similar mechanisms underlie various kinds of meditation, and the core of any meditation technique is to prevent awareness from focusing on actions, keeping it within the subconscious or the superconscious. In accordance with the universal structure of activity, there are three types of meditation, with the focus either on activity, or on operation, or oscillating between operation and activity. The first case corresponds to an unperturbed "stream of consciousness" so much praised by many religious schools.[172] The second type of mediation is characteristic of magic; over-concentration on the operation background can cause violent organic reactions, mystically

---

[171] Psychoanalysis is entirely based on such indirect techniques, with the only difference that a psychoanalytic study is focused on the unconscious, while ASC analysis is interested in the transitions between consciousness and the unconscious. This is much like the difference between the two kinds of the study of resonant ionization in atomic physics: one can either study the structure of ionization continuum or analyze the details of its formation as the result of interference between direct and autoionization-mediated transitions.

[172] Oriental teachings usually come to the mind in this respect. However, similar practices existed in Ancient Egypt and in the European culture; also, the idea of religious ecstasy is an essential component in the major branches of Christianity.

interpreted as the signs of "divinity". Oscillatory mediation can be the most controllable and stable type; nevertheless, in a stagnant form, it can trigger psychotic disorders.

## Psychotherapy

It is commonly known that people can get free from their troubles sharing them with the others. Basically, this is what makes psychotherapy efficient in many cases. In any activity, since it is directed to a definite product, people transform their inner motion into an external thing, which can exist independently of them. Converting a part of one's soul into an outer form, one puts internal strains out, which helps to relax and attenuate personal problems.

Originally, intuitive methods of psychotherapy simply involved a person in a group activity, radically erasing personal disturbances by annihilating any individuality. Later, such collective sessions became a part of religious rites and carnivals. Numerous amateur practitioners complemented the official modes of psychological rehabilitation. Development of capitalism stimulated professionalization of psychological knowledge and practices, in the same time leading to wide propagation of knowledge in simplified and vulgarized forms. Numerous schools (or, rather, businesses) advertised the means of psychological manipulation and self-control, most of them appealing to some esoteric mysteries pretending to be borrowed from deep antiquity and re-discovered today... Though many psychological and psychotherapeutic techniques got duly refined in that way, the trail of prejudice could never make both professional and vulgar practices efficient enough. One has to realize the universality of consciousness to be able to use it for better.

The well-developed hierarchical organization of activity should be considered as a cultural norm. Deviant activities can arise due to a number of reasons; they manifest themselves in various motivation disorders, distorted personality, or perverted mentality. Considering the universal hierarchy of activity, $O \rightarrow S \rightarrow P$, one concludes that the source of distress can be either organic (which is commonly associated with psychoses) or cultural (resulting in all kinds of neuroses). The most subtle cases of impaired subjectivity come from the combination of the two mechanisms; when certain organic inclinations (high sensitivity and reactivity) get enhanced by cultural restrictions (suppressed creativity), ideation disorder can develop.[173] Obviously, organic disorders do not disappear because of mere talking about them; similarly, communication cannot remove social problems. Any therapy can only soothe the pain, to let the person regain the reason, which is the only guide in really solving any problems.

Psychotherapy is an applied area based on psychological knowledge; however, psychology in general is not therapy oriented, and it does not at suggesting any techniques to protect the person from internal troubles.[174] Studying the motion of the soul has nothing in common with controlling this motion; for psychology, any outcome is equally acceptable, as soon as it does not violate the laws of inner dynamics. This position retains the psychic flexibility that is the necessary premise of subjectivity. The very universality of the subject invalidates attempts to reduce conscious behavior to a number of standard tricks, and the ideas about the "efficient" behavior (or success) depend on the cultural conditions, economic and social position of the person, education, nearest environment *etc*. Moreover, similar goals do not necessarily imply similarity of means. Multiple solutions exist in any situation, and individual choice is needed for every person in the specific environment. Different people cannot equally improve their performance (and health) using the same technique; to some extent some parallels could be traced in physiological reactions, discarding the influence of the higher (social) levels—still, in many cases, the higher (mental) processes essentially modify the background conditions for the lower-level (physiological) processes, and subtle organic interference can lead to most unusual results.[175]

The purpose of therapy is to support people rather than merely classify them. Tests could indicate how the person's activity and inner processes can be influenced, but they never suggest a choice. For instance, questionnaires processed using factor analysis can only give syncretic knowledge, which is not enough to predict or suggest anything. To overcome the shortcomings of psychological empiricism, one has to invoke a theory, and it is the quality of

---

[173] All psychotic disorders of unclear nature are traditionally put in the rubric "schizophrenia", and the only treatment known is to damp both neural motility and social extravagancy with heavy neuroleptics and brutality of psychiatric personnel.

[174] Scientific knowledge in general says nothing about its possible usage; the latter is determined by the economic and cultural factors, and no individual can change this by a however good intention.

[175] Without accounting for this possibility, medicine would reduce to mere veterinary. A person should not be identified with the organic body, and the ways of treatment must involve active co-operation of the patient and the physician.

the theory that determines the relevance of the conclusions drawn on the basis of a test. However, a good theory is not enough to change people's behavior and psychology; one needs philosophy to indicate the directions of development. Incorrect choice not accounting for the peculiarities of the current cultural formation can result in severe distortions of personality virtually dissolving any subjectivity. A good psychotherapist, in addition to practical tools and theoretical knowledge, must feel the tendencies of social development and discern their reflections in cultural phenomena, including philosophy. Since universality is rather limited in a class society, therapists cannot be expected to possess a solid philosophical background; they have to depend on some philosophy of consciousness. In the cultural formations to come, the very need in professional psychotherapy will be eliminated.

Psychological assistance is to indicate the possible ways of the person's (or group's) coexistence with the objectively formed cultural environment; when the people cannot change the society in a desirable way, they have to learn to tolerate it minimizing both individual and social harm. This can only be possible, if the person participates in a higher level activity directed to the reorganization of the society on the grounds of reason, towards more freedom and deeper spirituality. In other words, there must be sense in one's life, and the therapist is to help the patient to discover it. Though it is not in the competency of psychology to form a general view of the world, a good psychologist can teach people to consciously face a difficult situation or unfavorable position; freedom begins with the honest perception of the world and society, showing what can be done here and now to become common cultural heritage in the future. Efficiently using one's abilities towards the universal goals helps to avoid the destiny of a miserable career-maker manipulating the others and crippling one's self.

The inner organization of the subject depends on the overall way of life, and any psychological change is to come from some outer effect. Normally, a therapist cannot (and should not) change the patients' circumstances; this must be primarily achieved via conscious action of the patient, albeit supported by other people. Therapy can induce a process of active reassessment of the patient's cultural environment and internal restrictions and thus help in finding a satisfactory solution strengthening the universal components of the patient's life, rather than mere physiological adaptability. Obviously, analysis of sexual anomalies or cerebral dysfunction can hardly help in stimulating creativity and enhancing spirituality.

Technically, a therapist affects the patient's internal state creating an artificial cultural environment triggering desirable behavior,[176] with the degree of reflexivity enough to induce changes in inner motion. That is, a patient is involved in a joint activity unfolding the problematic sides of the personality. One cannot see the other's personality without making that other act and behave. However, in the psychological study, there is much more mutual influence than, say, in the case of a physicist interacting with a quantum system; the efficiency of therapy is directly related to the spiritual growth of the therapist in the course of treatment.

A good therapist would combine methods of different sciences and non-scientific reflection to obtain all the relevant information about the patient. Psychological tests are only a part of the available technologies, referring to very special sides of subjectivity; this is the analytical level, measuring the responses of a conscious system to a well-structured set of stimuli. On the syncretic level, to precede formal analysis, projective techniques are widely applicable, which is closer to art than science. For an integral understanding of the patient, a therapist would use formalized and informal categorial schemes sublimated from the scientific and aesthetic pictures of the human soul and spirit, as well as borrowed from philosophical tradition.

Generally, the dialog between the therapist and the patient unfolds depending on the social position and experience of the patient, and it is important to be aware of the usual conditions of the patient's existence. There are no rigid schemes applicable to every person in any case, no single model of the internal world that would be applicable to any personality; evaluating the problematic situation is far from putting the patient in a pre-defined structure. The larger is the therapist's acquaintance with theoretical views and practical models, the better; however, combination of different elements in therapy is not entirely eclectic, it must always retain the general idea of universality as a determinative characteristic of subjective mediation and consciousness. Otherwise, an integral view of the person is hardly attainable, and the discrepancies in reflection will lead to psychological contradictions and behavioral conflicts.

Attempts to "program" people, reducing their behavior to a collection of standard operations or psychological tricks, are bound to fail since the very idea of subjectivity implies transcending the limits of any activity, conversion of hierarchies and discovering new ways. One cannot measure the efficiency of therapy by the number of habits developed or skills acquired,

---

[176] Thus, psychological types in the psychotherapeutic sense are nothing but the classes of environment required to control the person's behavior.

since it says nothing about the actual impact of the techniques learned on the patient's life and activities. There are no absolute criteria of effectiveness; in a psychological approach, patient-centered methods and individualized conceptualizations are preferable. Thousands of possible actions serve the same purpose, and one has to choose those most consistent with the social demands and the peculiarities of the patient.[177] One can never say that the patient's psychology would evolve in the same way after the application of the same technique. This distinguishes psychotherapy from medicine, which is essentially based on the determinism of organic reaction on an external intervention; however, even in medicine, the transmutation of physiological phenomena due to the influence of higher-order psychological processes must often be taken into account to achieve the desired result.

Psychotherapy may formally resemble psychological manipulation, since it implies bringing a patient in a state selected by external to the patient criteria, often against the patient's predisposition, and sometimes by the way of rather painful experiences. However, psychotherapy is different from manipulation in that it is consciously initiated by the patient, rather than therapist; therapy is carried out in the interests of the patient. This also distinguishes psychotherapy from psychiatry: psychiatric treatment is mainly applied to a person causing social disturbance by inadequate behavior, and hence is governed by social rather than individual interests.

Psychotherapy does not teach people to manipulate the others; it is aimed at increasing people's ability to control their own behavior and mood. Yes, flexible behavior can help in adjusting the social environment to certain kinds of activities, but such influence should be achieved as a by-product, never being a motive (though, sometimes, it can become a transitory goal).

Since any changes in the subject can only be the products of certain activity, the central place in any psychotherapy is occupied by the category of *work* as opposed to *labor*. An activity can have a therapeutic effect only if it involves the patient as a subject, and the changes produced in the patient's environment and internal world have to be consciously accepted as one's own products. Work may be hard, but it is always creative and purposeful, and it grows from inside the subject rather that from the others' prescriptions or externally imposed needs. Abstract manipulations devoid of meaning and sense, without personal connotations and social reference, can never be psychologically helpful; lack of enthusiasm on the patient's side is quite

---

[177] Thus, the most effective way of fighting with cancer would be exterminating all the humankind with all its maladies; however, few doctors would recommend that radical therapy.

natural in such therapy.[178]

It is important that the patient and the therapist consciously collaborate, sharing their purposes. This unites them in a higher level (collective) subject, making the joint activity hierarchical. Without that, any therapeutic methods are bound to reduce to treating physiological (primarily sexual) problems rather than people's souls, with interaction of the therapist and the patient becoming a kind of sublimated sexual act aimed to a sublimated (or psychopathic) form of orgasm, be it tête-à-tête or in a group.

Since the idea of a collective subject can rarely be acceptable under capitalism, the therapeutic methods used by bourgeois psychologists are restricted to the aspects of subjectivity compatible with the norms of the capitalist society. Objective development of science virtually violates any social limitations being determined by social practice in general and hence by the zone of imminent development for the whole society rather than any subculture. Growth of the subject's universality is the main goal of efficient psychotherapy; however, this implies, in particular, becoming aware of the existing economic and social limitations, and gives a motive for global social reorganization. While such reorganization is economically premature, there is a problem of adaptation to the realities of life after the course of therapy, without transforming the acquired skills into mere manipulation. If one fails to universalize therapeutic experience, the only way to keep an acceptable psychological vitality would be through repeating sessions of therapy, of either the same or a different kind; this can develop into a kind of dependence similar to drug dependence, smoking or alcoholism—for yet another example, workoholism could be mentioned.

The paradigm of consciousness as the boundary of the subject in a cultural space (or a motivation space) as pictured in figure 5 implies dynamic aspects. The shifts of consciousness can be treated as changes in the shape of the boundary caused by the local balance of inner and outer pressure. Normally, inner pressure dominates, and the subject's tends to expand, assimilating new cultural areas. Under certain conditions, this expansion can encounter an obstacle, and the continuing expansion in the adjacent regions results in inner lacunas in the subject (figure 8). The boundary of such a vacuole in the subconscious as similar to the external boundary of the subject; the subconscious tries to reduce the size of the lacuna, but is unable to do that, only increasing the density of the "bubble" so that it can stand any

---

[178] The "therapist" practicing such an approach will always be able to attribute the failure of the treatment to the insufficient effort from the patient.

pressure. The presence of the super-conscious in the subconscious is felt as neurotic tension.

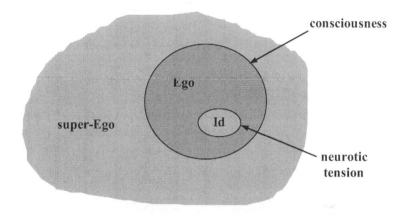

Figure 8. Lacunas in the subconscious as neurotic centers.

This picture suggests a number of important implications. First, no person can remove neurotic formations without the help of the others. The whole inner representation of the cultural space is required, which can only be achieved via activity exchange. Second, the only possibility to connect an inner lacuna with the rest of the superconscious within the same hierarchical structure is associated with cultural deprivation destroying part of one's subjectivity; this is what usually happens in clinical treatment. However, figure 8 indicated yet another direction of therapy, which is more appropriate for a conscious being. The disconnected portions of the inner space can be connected extending the dimensionality of the space, which is equivalent to spiritual development. That is, involving the patient in new activities, or extending the range of existing activities, a therapist can switch the main direction of personal growth to the new dimension, thus showing the way to inner integrity. The very therapeutic session is already extending the patient's experience; quite often, this is enough to break the inner barriers.

The application of every psychological technique assumes an appropriate level of communication. People simultaneously communicate via different channels; any inconsistency in the communicative positions is immediately detected by the partners, at least as a vague feeling. For instance, in the transactional model, the partner expecting transactions of the $C - A$ type will object to mere simulation of this type of transaction on the analytical level, which can be interpreted as humiliation, provoking a very sharp reaction. The analytical technique of approaching the $A-A$ scheme through the oscillation from $A-C$ to $C-A$ and back will be hardly applicable in this case. Conversely, when a series of analytical "strokes" is expected, exaggerating the $C-A$ transactions can be treated negatively.

Application of certain psychotherapeutic techniques may be prohibited by psychotic or neurotic limitations; for instance, the existence of a tough aversion area around some transaction schemes means unavailability of the psychological techniques based on such transactions.

The "wounded child" and "screwed parent" positions are among the typical examples. The former is a result of heavy social pressure demanding obeisance and dependence—a child is taught to "behave"; the latter is due to exaggerated responsibility, the persistent necessity to "get distinguished" and "prove superior". The $C$ component of a "wounded child" is stagnant, demanding care from the others; the $P$ component of a "screwed parent" cannot allow any doubt in the person's authority. A "wounded child" fights with real or imaginary attempts of external pressure with the purely childish means, that is, blind rebellion and spontaneous (self-)destruction; a "screwed parent" struggles against too much responsibility with the means of a "parent", striking attitudes, blaming the world, or demonstratively "washing hands".

Transactional impairments can manifest themselves only in syncretic communication. Thus, in many situations, a "wounded child" can play child without feeling and psychological discomfort. The problem is to include the deviant person in a wider hierarchy of interpersonal relations, with the focus on some other transaction types. In particular, since intimacy is impossible without a syncretic component, the presence of a stagnant position in one of the partners will make intimacy unstable; it can be broken with much pain if a wrong key has been touched by chance. This is how great friendship may break—or love may die.

Since meditative techniques require concentration on an object or inner state, their application is equally able to release stagnation in another area and provoke stagnation around the meditation center. The psychotherapeutic usability of mediation is based on its ability to unfold too rigid operations, thus giving them sense. However, the traditional meditation practices that initially developed within various religions, in general, are not therapy oriented: releasing inner tensions, they do not suggest new dimensions of personal development and hence imply infinite repetition—this betrays their religious origin, as religion is not interested in creativity, it needs addiction.

There is a universal technique that is as powerful as meditation in destroying inner stagnation areas, while devoid of the negative consequences of prolonged meditation. The conception of a neutral state in inner dynamics has been formulated by Hasay Aliyev in the end of XX century, and it is entirely compatible with the idea of hierarchical conversion, which requires first folding the hierarchy to an unstructured state, a "point", then unfolded in either direction. That is, to consciously change the direction of inner activity, an intermediate activity is needed that would not be related to both the initial and target activity. In a neutral state, there is no focus of awareness at all; awareness is spread between the levels of involvement. As with language, there are activities especially suited for such mediation; they can become a kind of universal mediator, thus acquiring subjective aspects. Psychological assistance in finding an individual hierarchy of neutral states produces most stable regulatory mechanisms without interfering with creativity and spiritual development. Wide acquaintance with such techniques is a step towards elimination of professional psychotherapy as such.

It is important for a person to have a range of socially admissible faults, a zone of acceptability; small behavioral deviations should not impair the person's image in the eyes of a certain reference group, and hence one can have more self-confidence, more room for self-construction. The reference group can either actually exist or be an imaginary construction representing a collection of dispersed social attitudes. Psychotherapy can influence the choice of the reference group and thus increase the level of self-confidence. However, in therapy, this can only be should not be a transitory goal; a malignant (or mercenary) manipulator can efficiently lower the criteria of self-assessment, thus disintegrating consciousness rather than supporting it.

If there is somebody who esteems a person in a syncretic manner, as a whole, rather than for certain isolated qualities, this person won't put much importance on temporarily occupying a lower position in communication with anybody else, even experiencing a strong "parental" pressing. Once self-esteem has formed as a reflection of external recognition, it opens the capability of expressing esteem for the others. Psychological training can increase one's sensibility to the signs of esteem from the others, and hence flexibility of conveying esteem back to them.

Formal analytical communication is the only way to remove internal strains: consciously designing transactions to externally model one's inner world is the basic mechanism of projecting that inner world out, making it distanced and abstracted from personal feelings. This approach is employed in most methods of psychotherapy and techniques of psychological regulation. However, formal results cannot pertain without support from outside; they rapidly vanish after the original communicational environment is no longer accessible. Consolidation of personality shift due to analytical training implies ability of transferring thus developed transactional habits to other social environments; that requires the "immersion" of an external habit into the structure of personality, a syncretic "fusion" with it, which is not possible without synthetic communication with good friends.

Since conscious behavior is not a result of neural or other physiological processes, any attempts to treat psychological problems with medication are wrong in their premise. Drugs, or induced organic reactions, can alter the person's behavior inasmuch it is biologically determined; since subjectivity is predominantly related to the non-organic body of the person, the conscious side of behavior can only be controlled through changes in the person's involvement in social processes, that is, creative work and productive communication. Similarly, no words can cure inner pain, since consciousness is impossible without the subconscious and superconscious levels, which are not influenced by verbal manipulations. Object-mediated communication is primary, as compared to verbal communication, and the objective links within the society dominate over individual transactions. People must live together and work together; they need more conscious experience and more directions of cultural development—this is the best therapy possible.

# ETHICS OF CONSCIOUSNESS

Once the subject has been understood as irreducible to a biological or physical body, and universality has been taken for the determinative feature of subjectivity, the obvious corollary is that an individual can only be a partial representative of the subject, necessarily combining subjectivity with the features of an animal, or inanimate body, and one has to consider the degree of subjectivity in every particular person or act. Thus, moving in space with the Earth, one behaves as a mere physical body, rather than a conscious being; similarly, the necessity of eating, or whether sensitivity, has been inherited from the biological ancestors of humans. A conscious act, beside its organic or physical implementation, is characterized by its cultural function, and hence its relations to the society in general, the collective subject. Reflected in the conscious individual, these relations take the form of awareness, intentionality and responsibility. In their unity, they give the subjective idea of the self.

Self-consciousness is the necessary level in the development of consciousness; the unity of consciousness and self-consciousness is the highest form of subjectivity, reason. The manifestations of reason in an individual's behavior must also be present in reflection; that is, along with the objective assessment and subjective opinion of every act, there is *ethical* evaluation, judging one's activity by the standards of reason, by compliance with the principal directions of cultural development. This judgment is in no way disparaging: it does not treat non-conscious behavior as inferior to conscious activity; it simply states the distinction of the subject from physical bodies or organisms. Moreover, ethical judgment indicates the presence of the spirit in every physical, organic and behavioral process involved in conscious activity; this sublimates lower levels of reflection, giving them sense.

The physical and organic modes of mediation are marked in the subject with a specific subjective quality. Doing apparently the same as the animals, a conscious being will do that in a different inner and outer environment, modifying and transforming mere physiological reactions into something quite different. Thus, a human can eat, drink, or sleep like any animal—but these simple actions are performed differently by humans, and even when the necessity drives a human to lick water like a dog, this is only a temporary mode of action that will give way to other forms in a different situation, unlike the pre-determined reactions of an animal.

There is no unbridgeable abyss between conscious and non-conscious life; one can find rather complex modes of behavior in animals, as well as very primitive operations in humans. It is only in the society that a human can be considered truly conscious, while all the organic functions can as well be reproduced by specially conditioned animals. It is important to consider the development of the inner hierarchies in the subject, to be able to discern subjectivity in animal-like behavior, or conversely, discover mere complex reflexes under seemingly conscious acts. This ability is related to both discerning subjectivity in the others and self-evaluation. It is objectively necessary, and the level of ethical judgment is a measure of subjectivity and consciousness, along with logic and imagination. Observing the conscious character of the other's activity (respecting universality in the others), one becomes conscious as well.

However, the very universality of mediation implies the presence of lower forms in higher activities. No person can be conscious in any respect at any moment. A conscious act is always accompanied with many animal-like operations, as well as various physical or chemical processes. In integral behavior, such lower levels remain under conscious control as long as one can choose between different alternatives, arbitrarily switching from one mode of action to another, according to the turns of the situation. If an individual is put in the conditions that regularly demand pre-determined response rather than conscious choice, conscious acts degenerate into mere reflexes.[179]

The controlled shift of identification from subjective identity to various objects is a mechanism of cultural growth; in the subject's inner world, it becomes reflected as conscience, which is a universal mechanism of self-regulation. Assuming universality in the others, implies deep esteem of any manifestation of consciousness.

---

[179] In the states of objectively reduced consciousness (*e.g.* under alcoholic intoxication, or in slavery), humans become animals rather than conscious beings.

## Syncretic Ethics

The definition of the subject as universal mediation provides clear criteria of subjectivity that can be applied to any individual or group in any situation. To be related to consciousness, some individualized formation must, first, be mediating the relations between objects, and second, do it in a universal way.

### *Mediation*

To be conscious, one primarily has to operate with objects producing other objects. No consciousness can be born within the soul; without outer action consciousness cannot grow into self consciousness and reason. Our thoughts and feelings are mere reflections of our subjectivity rather that subjectivity itself. Primarily, we are, and we are in the world, and our self-reflection is a component of our reflection of the world.[180] The growth of consciousness is mediated by productive activity; however, the products can be of rather different nature, and they are reflected in different hierarchical positions, including those with the object's ideality on the topmost level, which may look like entirely ideal activity that does not require material things at all. Of course, this is sheer illusion, since the most abstract thought is virtually a physical motion, albeit hierarchically organized and correlated.

It is not always easy to decide, whether the person performs mediation or occupies the object/product position. People are often manipulated by other people, thus losing their subjectivity in that respect; as they remain subjects in some other activities, the overall impression of conscious behavior hinders detection of hidden dependences. This is how political indoctrination works, this is the basis of neuro-linguistic programming and any swindler skills.

Playing the role of an object or a product is normal for any subject. This is necessary for the subject's universality, since one's position in any activity is a matter of conscious choice as well. Shifts of consciousness accompany the process of activity's conversion, implying temporary "neutralization" of consciousness. As long as one keeps control of that balance of subjectivity and non-subjectivity on a higher level of hierarchy, the overall activity

---

[180] The Cartesian principle *"cogito ergo sum"* is still valid, since our ability to think is based on our existence; the inverse is not true: there are aspects of our existence that are not related to thought.

remains conscious. It is only when social conditions prevent free conversion of hierarchies that the object or product positions can lose their deliberate character, with impaired subjectivity.

The subject's self-reproduction in the $S \to O \to S'$ cycle is to keep the mediating position, always implying the primary cycle $O \to S \to O'$:

$$O \to (S \to O \to S') \to O'$$

Doing something for themselves, people mean something to be done in the world, some cultural effect. Reproduction of the subject is impossible without reproduction of the cultural environment, a hierarchy of objective mediations. Restricted to inner structures and processes, reproduction will lose its subjective character, becoming reproduction of an object rather than conscious self-reconstruction. In particular, mere survival behavior is devoid of subjectivity, and egoistic motives are incompatible with reason.

## *Universality*

It would be easy to decide on the universal character of anybody's behavior, if we could have observed its influence on the world as a whole, including its past and its future. Still, this is utterly impossible, simply because it would contradict to the very definition of a finite person, a finite observer and a finite behavioral act. Living in the world, we have to guess about our place in its integrity, and this is what reason is given us for.

### *Objective universality*

Universal mediation is to be implemented in non-universal things. The material carriers of consciousness (objects, or collections of objects) must be complex and flexible, performing mediation in a strongly coupled mode, to implement the unity of reason. There are infinitely many requirements to the possible material implementations of consciousness, since infinity essentially enters the definition of subjectivity. The hierarchical approach explains how all that complexity could be contained in a finite object.

In hierarchies, the very distinction between the finite and the infinite is relative. Every object is a hierarchy that can be unfolded in infinitely many positions, and every two levels of that hierarchy can be connected via infinity of intermediate levels. Objectivity implies the unity of matter and reflection, the material and ideal aspects of the same reality. While the material side of the object determines its being bounded on every fixed level of hierarchy, the ideality of the object is free from any spatial and temporal limitations, containing an *actual infinity*.[181] The reality of the object is a synthesis of its finite existence and infinity, their mutual penetration and transformation into each other, which could be, to certain extent, related to the idea of *potential infinity*. Reflecting an object in a universal way, we reflect its material and ideal aspects, and hence comprehend both its finitude and infinity.

The objective side of the subject's universality is subjectively reflected as the problem of one's trace in the world. Every person's activity, since it contains the conscious component, must be objectively represented (the product of activity), and this imprint is not subject to death, being one's unique contribution into the world's development. However small that contribution may seem, it is indispensable for integrity of the world, and thus the smallest trace acquires cosmic significance.[182] The problems with self-esteem, or self-respect, is not, therefore, related to the negligibility of one's existence, but rather to people's inability to comprehend their real value for the world; to some extent, this problem can be solved by stimulating creative reflection to discover the universal sides of common activities—however, in general, it requires a social mechanism of linking one's consciousness to its objective roots, discovering the sense of life.

*Subjective universality*

The universality of the subject means personal freedom. This freedom covers one's thoughts and actions, and it can manifest itself as relatively

---

[181] This delicate question has provoked much argument in physics. For instance, if two spatially localized particles interact, where is the energy of their interaction? Where is a photon? Can the temperature of an atom be defined? It is usual in physics to attribute such properties to some "field" occupying all the space, or even space-time. However, this provokes new questions, which are not so easy to answer. Thus, if a photon is (spatially and temporally infinite) electromagnetic field, how can we speak of its emission or absorption? If light propagates at a finite velocity, why do we speak of the field in the spatial regions, where the photon has not yet arrived? In quantum physics, every two electrons are considered as entirely interchangeable no matter how far from each other they may be! Situation is exactly the same in considering people's activity and psychology: every two individuals are tied together, even being entirely unaware of each other's existence.

[182] In many religions, the idea of a god is associated with the power of creation, up to the creation of the whole world. Leaving material (and cultural) traces, one places oneself "among the gods"—metaphorically speaking, creative people "join the gods" after their bodily death. The opposite of the "godly" (creative) side of one's existence could be metaphorically called "satanic"—this is the destructive element. The both aspects coexist in the same activity, and one's cultural heritage may be regarded as either positive ("heaven") or negative ("hell") in different respects or in different epochs; the truth, as always, is in the synthesis of the both.

independent of environmental restrictions. Without freedom one cannot attain spiritual integrity, and hence be truly conscious.

The subjective aspect of behavioral universality is closely related to the problem of behavioral styles. Conscious behavior is universal not only by its content, but also by its form. Therefore, the presence of a certain type of behavior within any given cultural formation assumes a specific level of universal mediation, determining the subject's position in any activity. The styles of conscious behavior are both shaped and manifested either in special activities, or as a result of conscious action, or even as an outcome of an operation. Behavioral styles may resemble behavioral habits; still, they are qualitatively different from mere individual patterns observed in animals, since the formation of a style is essentially its production, hence involving consciousness on all the levels. Behavioral styles are a part of the gross product of the society as a collective subject; once produced, they become an element of the culture.

Styles of behavior should not be confused with their mechanisms. Universal schemes of activity still have to be implemented in a non-universal complexes of habits, involving physical or physiological processes; however, styles also contain a cultural component, a special organization of lower level elements that cannot be deduced from their own dynamics.

*Cultural universality*

The universality of the subject as a product of social development is reflected in the category of *spirituality*. Spirituality is a level of the culture originating from cultural experience and growing into praxis.

Syncretic spirituality contains such forms as religions and common moral; the non-critical character of this level may make it extremely dogmatic and thus alien to subjectivity itself; however, in a due proportion, such syncretism is necessary to maintain the stability of the society and to link the spirit to material culture. The already achieved must be preserved as the basis for further development—provided the higher priority of development is ensured.

Analytical spirituality is a cultural representation of creativity as a necessary component of any conscious activity; in the institutionalized forms, creativity is made the purpose of activity rather than its mechanism. The three fundamental levels of creativity are art, science and philosophy.

Institutionalized spirituality becomes common experience, objectified and assimilated on the higher levels of activity. In this way, every type of spirituality becomes hierarchical, reflecting the other levels as its internal motion. Due to the universality of subjective reflection, every act of a person as a conscious being is saturated with spirituality, and the spiritual load of an act becomes the measure of its conscious component.

However, abstract creativity cannot yet make a person creative; true creator will primarily redesign the world, rather than exteriorize one's inner motion. The highest level of cultural reflection, praxis, is the synthesis of simple reproduction and spirituality; this is creative work consciously transforming both nature and spirit in accordance with the objective directions of development. Thus understood work is different from mere effort caused by physical or physiological necessity; it is the opposite of mere play, whimsicality, or labor, which often reveal lack of spirituality. Work is driven by one's own needs as they coincide with the needs on the world. Once one makes the world the product of one's work, this opens perspectives for self-reconstruction as the unity of the material and ideal aspects. This implies all-encompassing interdependence, when everybody is responsible for everything, being dependent on everyone, as well as the impossibility of one's subordination to any other person. Work makes irrelevant family, clan or class affiliation, economic or political conditions, religion or moral norms. Virtually, one becomes free of being enslaved by oneself.

Cultural formations based on the division of labor are marked by limited sociality. In such societies, freedom can only partial, inconsistent, hidden in singular deeds. Economic development is bound to replace the present economic and social system by another formation that will be more suited for universal spiritual development. Still any conscious act can only be judged from the position of the embedding cultural formation, and we have to show reason today, looking to the deeds of our ancestors with their own eyes.

## Analytical Ethics

The subject is universal mediation. This simple formula can be unfolded in a hierarchy of special principles that admits many conversions, and all the resulting ethical systems are equally admissible. Departing from the universal structure of activity, $O \to S \to P$, one would consider three factors:

1. The objective aspect of consciousness is related to the adequacy of

reflection and hence the level of self-control; this aspect is referred to as *awareness*.

2. From the subjective angle, consciousness requires behavioral consistency related to the integrity of the self, the *identity* of the subject; in ethics, this level is primarily related to *intentionality*.

3. The productive side of consciousness is related to free will, and hence one's *responsibility*.

All the three components are necessary for a truly conscious act.[183] If one of them is missing, behavior becomes "deviant", or even "insane". For instance, "field behavior" (or "impulsiveness") is characterized by high awareness, but the absence of the subjective focus makes it too chaotic for consciousness; this is especially characteristic of small children, with proper responsibility yet to be developed. The opposite situation, lack of awareness along with well developed identity and responsibility, can lead to neurotic or psychotic deviations; psychotherapy extending the range of awareness is advisable in this case. There are also examples of weak identity accompanied by both awareness and responsibility—ranging from avoiding one's dentist to split personality. With all the possible combinations, such situations represent the different cases of impaired universality, and hence lack of consciousness.

### *Awareness*

Awareness is a component of consciousness that reveals the kinship of humans to the rest of life, their animal origin. Higher animals seem to possess syncretic awareness, the ability to distinguish biologically important events and react with specific mobilization. When animals live with people, this conscious environment modifies their behavior, synchronizing their physiological processes with human activities; this often would enhance pre-conscious awareness, and many domestic animals demonstrate very sensitive behavior comparable to that of human children. However, human awareness is different in that it relates the sensations of an individual to the social norms

---

[183] Similar criteria of consciousness are traditionally used by the common sense. Also, in a legal inquiry, a person cannot be convicted for a crime if the person acted without knowing it, or suffered from a mental disease, or acted under external persuasion, being forced to commit the crime against one's will.

and attitudes.[184] A conscious person will not only observe the occurrence of some events, but also detect their social origin or predict cultural consequences.

This results in a generalized subjective picture of the world, neutralizing individuality in favor of universal cultural representations. The "portability" of conscious perception allows operating with socially mediated reflections of real events, facts. Such awareness can be communicated from one person to another. In this shared existence, it becomes interpersonal, collective awareness, knowledge.

Self-awareness is a special case of awareness directed to the subject's own activity. In particular, a conscious individual is aware of his or her actions or feelings (outer and inner activity), and this self-reflection is as generalized as awareness of external things and their relations: a person can only be self-aware in a definite social role, as an instance of cultural position. Even those who feel themselves absolutely unique and expelled from the society are paradoxically representing uniqueness and alienation as cultural functions.

People are not always aware of the world around them and their own acts. However, this lack of awareness may be intrinsically related to consciousness, only driven to its periphery by the objective situation. Such subconscious and superconscious layers of awareness can usually be drawn to consciousness in special activities, and they constitute a part of memory and experience. To happen outside the person's consciousness, something must be completely unrelated to one's social existence. Thus, much of our physical interactions with the world, or tissue metabolism, will never come to consciousness unless it happens to hinder one's ability to consciously act. People move and dream in a mechanical way, their muscles work, neurons fire—but this is only a natural background for conscious acts, coordinating the same physical and organic processes through communication with the other humans involved in a common activity.

Wakefulness is related to awareness in that it activates the body allowing its immediate involvement in conscious activities. However, one does not need to be awake to become aware of the inner and outer activities, at least at the periphery of consciousness. One can be aware of one's dreams, and some creative people can work while sleeping, though they usually have to wake

---

[184] In dialectical materialism, consciousness (*das Bewußtsein*) is primarily treated as awareness of one's being (*das bewußte Sein*), and one's way of life virtually determines one's consciousness.

up to register the results of such work and make it socially available.

Awareness is hierarchical, and people can act as both conscious and nonconscious beings in the same time. For instance, they can be aware of what they are doing here and now, but unaware of the distant social consequences of their acts. Similarly, deception makes people unaware of what they are really doing, which means that their behavior is only partially conscious.

## *Intentionality*

Awareness, forming the core of consciousness, is not enough for truly conscious behavior. A conscious being does not merely follow the course of inner and outer events, but attempts to direct them to a definite result. The distinctive feature of conscious behavior is that its outcome ideally exists before the actual implementation, and this imaginary result can coordinate the efforts of many people, uniting them in a single organism, a collective subject. A conscious person's action is always directed to a definite goal; even when people deliberately relax and do nothing, they pursue a quite definite purpose of relaxation and leisure. Conscious beings are always producing something, and these products are intended to be used by other conscious beings (and reflexively used in particular).

Since any ideal formation is a relation between material bodies, human goals exist as hierarchies of material things as well. Humans have developed numerous mechanisms of establishing subjective goals as objective social structures; the hierarchy of art, science and philosophy is an example of such objectified expression.

Conscious intentionality is different from mere directedness of animal behavior to an outer thing, though there is a range of intermediate forms between the two. Consciousness implies the ability to make one's goals available to other people as their common goals. Already in animals, a biologically significant stimulus can coordinate their physiological processes (including brain activation patterns) to achieve a level of behavioral integrity much above mere spontaneity. In humans such external coordination becomes the dominant regulative principle, and one's behavior can only be called conscious inasmuch as it is controlled by a socially established goal.

Though both awareness and intentionality are the attributes of conscious behavior, they do not necessarily imply each other. One can be unaware of one's intentions, or intend something one is not aware of. Quite often the true goals of an action become known only in the end of it, and people can pursue one goal being aware of another. Human behavior is only conscious to a certain extent; however, the conscious component grows with the humanity moving farther from the animal world.

*Responsibility*

Doing something under the pressure of circumstances, without choice or control, is hardly compatible with the idea of consciousness, even if such actions are intentional and sentient. Less freedom means less consciousness. Responsibility as the other side on free will is intimately related to the social origin of consciousness as such.

Apparently, freedom is the ability to act following one's own desires and needs, despite the objective obstacles and external prescriptions, deliberately choosing from multiple possibilities. Sometimes, people seem to act without real necessity, as if just exercising their free will. However, such freedom has nothing to do with consciousness unless it is complemented with responsibility, which represents the social necessity of the behavioral act; that reflective free will is the only way for a physically bounded (finite) individual to act in a universal (infinite) manner.

Still, responsibility as a component of consciousness is more than mere external constraint on one's actions. Conscious responsibility is directed from inside out, and it is rather an internal restriction imposed by a person on him- or herself, that is, self-restriction. Free people *may* do anything, but they *will not* do many things because they won't want to. This self-control reflects the current cultural tendencies; in certain cultural conditions, it can be internally felt as external pressure, a kind of censorship not allowing the person to become truly free. To remove the psychological conflict thus produced, it must be transformed to productive behavior.

In a well-developed society constituting a collective subject rather than mere collection of individuals, there are special mechanisms to transform external restrictions into self-limitation, the practice of self-control becoming a part of the culture. Hierarchical conversion is the attribute of any activity; it must be redistributable among various groups of individuals, thus changing their social roles and responsibility.

Freedom does not mean the absence of any restrictions at all, since it is the cultural environment that unfolds an activity in a particular manner, thus forming actions with their conscious goals. On the other hand, activity can be folded and interiorized; the resulting flexibility of goals is the basis of the feeling of freedom and the psychological phenomenon of will.

The very necessity of freedom for a conscious being is a restriction on one's behavior. Socially inadequate behavioral acts are felt as indecent, more appropriate for an animal rather than human. In the cultures that are not yet hierarchical enough, one has to work hard to remain free despite the cultural restrictions; creativity may become a kind of obsession in the eyes of the others.[185] Better social organization will make transformation of freedom into responsibility a normal function of consciousness, with external freedom implying inner responsibility, and external restrictions being the basis of inner freedom. The both aspects get synthesized in subjectivity understood as universal mediation: as the necessity of reasonable action, it is an external (objective) restriction; as a product of one's own activity, it is a manifestation of free will. This resembles the basic mechanism of conscious self-control $(R \Rightarrow C \to R)$.

Animals differ from the subject in this respect. They try to satisfy their needs with a bodily reaction entirely controlled by the need; animals are easily distracted from the original direction by any concurrent need. On the contrary, a conscious person, to achieve some goal, proceeds by *organizing the environment* so that it would be impossible for the body (organic or non-organic) to behave otherwise; that is why conscious goals are much more stable. Objectification of one's intentions makes them conscious goals, imposed as external necessity; perception of the goals as one's own product is the core of free will; responsibility synthesizes the two processes in a kind of development cycle, establishing the unity of the subject with the world.

Responsibility is reflection of the subject's universality in the individual, and hence it must be a result of universal reflection, a product of some activity. Partially, it can be developed in properly organized education process. However, to become an expression of free will, responsibility must be consciously produced rather than imposed; it requires self-cultivation.

The lowest level of responsibility implies mere involvement in an activity, carrying out conscious actions within its scope. Once something is intentionally done, there is a conscious reason for the result—such *individual responsibility* could be compared to causation, with the production process in place of the natural concatenation of events. The next level of responsibility is related to self-consciousness; this is responsibility for oneself, or *personal responsibility*. Here, one becomes responsible not only for the results of

---

[185] It has become a banality to say that a genius is a deviation from normality, or even a mental insanity.

one's actions, but also for one's goals, purposes and intentions, as well as for the modes of action chosen. On the highest level, one finds *social responsibility*, encompassing also responsibility for the actions of the others—even for biological and physical processes involved in one's activity. This implies a synthesis of two complementary attitudes to one's environment: care and exactingness. A conscious agent will care for (and be careful with) everything, from the closest friends to the dust on the road.[186] However, the same agent will demanded from the environment to match the universality of the subject, and adjust itself to the subject's global predestination.

Subjectivity is hierarchical. In a collective subject, many individual wills seem to randomly combine in some residual will. However, this combination is not random; it is always driven by objective necessity, coordinating individual efforts in a common activity directed to a common purpose. Individuals are free within this hierarchy, acting as its representatives. That is, they are the subjects of their individual activities constituting different aspects of the embedding activity. This implies full responsibility: one either acts along the objective line, or rejects it; the degree of comprehension of the common goal is the measure of individual freedom. The hierarchy of one's will is built of the entirety of higher level goals, representing all the common activities of the individual.

Each individual is important for the integrity of the society in general; this is an immediate consequence of the universality of conscious reflection, which makes every member of society a representative of the whole, in a specific aspect. All people are equal as the carriers of subjectivity, and the history of the humanity is a result of a hierarchy of individual wills.

Responsibility supports preservation of the cultural heritage, extending the sphere of culture through re-creation of nature. When one's creativity opens wider horizons to other people, it makes their activities more universal, thus contributing to the development of subjectivity as such. On the contrary, any destructive actions hindering people's creativity restrict their power of self-expression and narrow their consciousness. A conscious being will care for other conscious beings and struggle against restricting their freedom.[187]

---

[186] "Tu deviens responsable pour toujours de ce que tu as apprivoisé"—but the destiny of the subject is to "tame" the whole Universe…

[187] Limiting the freedom of the other, you limit your own freedom; masters are no better than slaves. However, this applies only to the aspects of people's existence related to subjectivity: if a person does not act as a conscious being and thus restricts the universality of the others, such person must be restricted by the society, to prevent social harm.

Conscious responsibility is essentially supporting the traces of consciousness and reducing the animal and non-animate components in people's behavior.

Lack of responsibility may be due to either individual peculiarities or the social conditions. In the former case, the level of cultural development prevents an individual from becoming truly conscious, also limiting one's awareness and mutilating intentionality. In a rich cultural environment, there may be enough opportunities for education, while the social organization does not always favor too much creativity, and hence there is no social need in fully-formed subjectivity, in conformance with the level of economic development. This leads to both psychological and social conflicts; new aspects of consciousness can only be unfolded, when novel directions of activity are opened, with the corresponding reorganization of the society.

## Synthetic Ethics

On the synthetic level, all the aspects of ethical judgment become lifted up in an integral active attitude. Directed towards the subject, it becomes *conscience*; the same attitude towards the others is known as *esteem*. As an objective consequence of universality, the subject acting in accordance with the principal line of the world's development becomes immortal.

### *Conscience and esteem*

Since consciously acting people are to maintain the universality of their behavior, this ability becomes implemented in a special inner activity, which is subjectively felt as conscience. This is an essentially subjective formation, requiring a high level of socialization and freedom.

Originally, in primitive societies, there was no conscience. Normally, people simply identified themselves with their specific roles and hence were not responsible for any results. No guilt or moral obligations can exist in such a syncretic culture. People do something just because they have to.[188] It is only with the development of individual consciousness and self-identification that an inner regulator of behavior became available.

---

[188] Similarly, children acquire the ideas of guilt or moral obligation only after they attain enough social experience.

The appearance of such inner mechanism means shift of the reference point from the organism to environment. Animals always assess the situation from the same reference point, their body; they cannot put themselves in any other position. The formation of consciousness in the early people was possible since they could identify themselves with anything at all, including plants, animals, natural phenomena *etc.*[189] However, such identification was never arbitrarily, being directed by the tribal hierarchy. When there are few relatively simple activities within a stable hierarchical structure, their agents are easily identifiable with social roles. Later on, as division of labor grew, people had to combine different roles, which required labile identification. The level of economic development incorporated the necessity of choice. This was the origin of self-consciousness and the premise of individual morality[190] that could be quite different from the socially dominating morals, providing a personal frame of reference and enabling conscience as judgment in respect to that frame.

Being essentially social, conscience cannot develop as a generalization of punishment. Animal reflexes cannot be a starting point for the development of conscience; its origin is entirely cultural. Formal acceptance of the externally existing regulatory norms that leads to the feeling of guilt and expectation of punishment is not compatible with the subject's universality. The subject must consciously build its consciousness through social action and comprehension. There are no inherent morals; and one's conscience grows with one's subjectivity.

People communicate in many ways. Every communication act carries both objective content, related to common activity, and subjective content, reflecting the positions of the partners within the same activity. This activity provides the common categorial frame to make individual actions comparable and hence producing personal sentiments and attitudes. If one person finds no sense in the actions of another, no ethical judgment is possible. However, the subject's universality implies that any incompatibility of activities can only be transitory; as soon as different subjects get in touch, they will necessarily initiate an embedding activity uniting them in a higher level subject. In some cases, this process can assume the explicit form of learning more about each other, getting acquainted. There are also indirect

---

[189] The traces of this universal identification have remained in poetry.

[190] Play is very important for children as training in identification. It is utterly necessary for the development of self-consciousness; lack of play in the childhood leads to retarded personal development, and lack of conscience.

modes of mutual reflection, resulting in a variety of possible principles of ethical evaluation. These rules change from one culture to another, being always based on the universal mechanisms of activity.

One of the most important features of a conscious being is ability to discern consciousness in the others, which is reflected in the ethical category of *esteem*. Those who possess reason will esteem any other manifestation of reason, however different from their own. In particular, conscious activity carefully avoids restricting the freedom of the others. Of course, the very co-existence of different instantiations of consciousness implies mutual constraints, but these will rather be self-limitation due to universality of reflection.

Esteem is necessarily very exacting. One can be esteemed only to the extent of one's subjectivity. A conscious being is expected to properly behave, respecting freedom and demanding responsibility. Everybody may do anything at any time—but, if one's behavior becomes disturbing for the others or disastrous for the humanity, it cannot be considered as truly human and cannot be esteemed. Compliance with the idea of universal reflectivity and the correspondence to the main direction of cultural development is the principal criterion here. Since no finite system can be absolutely universal, no individual is perfect, and nobody can identify universality without mistake. It may take years to feel somebody's grandeur and guess a supreme destiny in a dull existence. People have to develop a kind of intuition for universality, to become sensitive to hidden signs of reason discerned by syncretic clues.

Respectful behavior is different from mere adherence to common moral or observation of somebody's rights. The highest morality may violate any moral norm or law if this violation is objectively justified as necessary for more universality; this is the necessary aspect of freedom[191] and the basis of self-esteem.

## *Immortality*

Animals are here and now. In humans, the formation of the level of conscious action expands the present moment both to the past and to the future, as the subject expands in the cultural space. The subject's universality

---

[191] However, such neglect should never degenerate into random spontaneity, scorn or disdain.

Synthetic Ethics

demands relating things in time as well; moreover, the subject can virtually reconstruct the very hierarchy of time, shifting from one scale to another. At some point, when the whole past becomes a part of the culture, people become immortal.

The idea of immortality accompanies the humanity from the earliest stages of its history; this indicates its inherent relation to subjectivity as such. However, primitive people were significantly dependent on their biological bodies, which lead to the limited picture of immortality as conservation of the physical body. When the human culture has grown enough, the idea of bodily immortality gave way to as abstract idea of spiritual immortality, reflecting the subject's independence of the organic level in the traditional religious form. After the discovery of the non-organic body of the subject (K. Marx), the true meaning of immortality could be explained.[192]

The destination of the subject is to transform nature to culture, to consciously reproduce the whole world, arranging it in a subject-mediated manner. Once some part of the world has become included in the culture, it has acquired universal significance. The products of conscious activity form a necessary level of the world, hence staying forever.

The hierarchy of the culture can alter its appearance: some regions can be folded and some other domains unfolded; the order of things may change. However, due to universal reflection, the folded parts of the hierarchy are always represented in all the other parts; in certain situations they can come back to the topmost level of hierarchy or, at least, form a distinct level.

Assimilating any part of the culture, one includes its universal content into the subject, thus extending the actual domain of universality. On the other hand, producing a trace in the world (and an impact on the society first of all), the subject contributes to universal creativity, allowing the whole culture to attain a higher level of development. When the same element of the culture is included in different domains of individual development, it becomes modified in a permanent way, including the history of its assimilation by different individuals into its inner hierarchy. Therefore, every activity is bound to produce everlasting traces in the world, and hence immortality of its subject.

---

[192] Today, the preservation of a biological body does not seem a good idea. People want to improve their bodies; they change their appearance and extend biological functionality using artificial tools and instruments. Virtually, people will be able to deliberately change the whole biological body, freely moving from one organism to another; more likely, they would leave their biological bodies aside and materialize entirely in artificial constructions.

The non-organic body of the subject consists of all the products of conscious activity; it is a part of the culture and therefore stays forever. The physical bodies implementing the non-organic body of the subject can be destroyed—this does not influence the hierarchy of cultural relations they represent, which will always find a different implementation. In particular, the older components can be restored as they were. This reproduces the idea of bodily immortality on a higher level.

The hierarchical idea of immortality is not as unusual as it may seem. It is quite common that a person occupies some place in the activities of the relatives and friends after the physical death, influencing their behavior and thus remaining an active personality. Some people make huge imprint on the history of the nation, or the humanity in general. Great artists, thinkers and philosophers continue influencing the minds long after their death; modern authors may often enter polemic with the authors of the past, as if the latter were still alive. On the other hand, there are personalities that have never existed in a biological body, being an entirely cultural phenomenon from the very beginning (literary characters, group authors, roles in online games *etc*). Recently, computer systems are beginning to demonstrate definite signs of personalization, producing virtual characters on the basis of the cultural positions of the users. We are getting used to virtual existence, and we could probably get used to immortality.